建筑立场系列丛书

新旧熔融
Additions: Integrating Old and New

[英]赫斯维克建筑事务所等 | 编

曹麟 邵磊 | 译

大连理工大学出版社

004 信息时代的建筑
在大数据和计算无处不在的时代，城市和建筑如何发生改变 _ Silvio Carta

挪威的风景线

006 挪威的风景线 _ Per Ritzler

012 埃尔德胡斯亚 _ Ghilardi+Hellsten Arkitekter AS

024 斯坦达尔斯弗森瀑布 _ Jarmund/Vigsnæs AS Arkitekter MNAL

032 Utsikten休息区 _ CODE: Architecture AS

046 阿尔曼纳朱韦特锌矿博物馆 _ Atelier Peter Zumthor

新旧熔融

062 新旧熔融 _ Isabel Potworowski

070 布里斯托尔老维克剧院 _ Haworth Tompkins

084 乌托邦——图书馆和表演艺术学院 _ KAAN Architecten

100 Musis Sacrum音乐厅 _ van Dongen-Koschuch

118 Coal Drops Yard购物中心 _ Heatherwick Studio

134 红十字会志愿者之家 _ COBE

144 普拉托新Pecci当代艺术中心 _ NIO Architecten

城市住宅新高度

158 城市住宅新高度 _ Heidi Saarinen

166 北塔 _ Reinier de Graaf/OMA

178 79＆PARK公寓楼 _ BIG

194 伦敦的储气罐 _ WilkinsonEyre Architects

206 Novetredici住宅综合楼 _ Cino Zucchi Architetti

214 Loftwonen 61号楼 _ Architecten|en|en

222 建筑师索引

004   Architecture of Information
       How cities and architecture are changing in the age of big data and ubiquitous computing _ Silvio Carta

Norwegian Scenic Routes

006   Norwegian Scenic Routes _ Per Ritzler

012   Eldhusøya _ Ghilardi+Hellsten Arkitekter AS

024   Steinsdalsfossen Waterfall _ Jarmund/Vigsnæs AS Arkitekter MNAL

032   Utsikten _ CODE: Architecture AS

046   Allmannajuvet Zinc Mine Museum _ Atelier Peter Zumthor

Additions: Integrating Old and New

062   Additions: Integrating Old and New _ Isabel Potworowski

070   Bristol Old Vic _ Haworth Tompkins

084   Utopia – Library and Academy for Performing Arts _ KAAN Architecten

100   Musis Sacrum _ van Dongen-Koschuch

118   Coal Drops Yard _ Heatherwick Studio

134   Red Cross Volunteer House _ COBE

144   New Pecci Center for Contemporary Art in Prato _ NIO Architecten

New Heights in Urban Housing

158   New Heights in Urban Housing _ Heidi Saarinen

166   Norra Tornen _ Reinier de Graaf / OMA

178   79 & PARK _ BIG

194   Gasholders London _ WilkinsonEyre Architects

206   "Novetredici" Residential Complex _ Cino Zucchi Architetti

214   Loftwonen Block 61 _ Architecten|en|en

222   Index

信息时代的建筑
Architecture of Information

在大数据和计算无处不在的时代，城市和建筑如何发生改变
How cities and architecture are changing in the age of big data and ubiquitous computing

Silvio Carta

随着无处不在的计算和互联移动设备的出现，城市正在发生巨大的改变。建筑越来越多地由真实的建筑构件（例如，墙、屋顶和立面）和非物质的部分（数据、互联网连接、无线信号、GPS导航和许多其他无线技术）的复杂组合构成。

也许城市在其基础设施中使用数据比在建筑中使用更加广泛。例如，通过城市仪表监控，我们几乎可以实时地对我们的城市有一个总体了解。在任何时候，市民都可以了解城市某一区域的污染程度、出租车和公共汽车的确切位置，或者了解特定的公共或私人事件的相关信息。

此外，使用城市数据可以对城市的情况进行更为广泛和深入的考虑。这包括某些社区的富裕水平、特定地区的安全指数或公共交通的总体质量。今天，市民们可以更加了解他们所居住的城市是如何运作的，以及它随着时间是如何变化和演变的。人们现在可以看到他们城市生活的某些方面，在数字时代来临之前这些方面都是隐藏的或者是难以解释的。今天，成为我们城市特色的数字基础设施不仅为人们提供了对周围发生事件的新的关注点，而且更有趣的是，让他们对自己的公共生活和城市生活产生了新的理解。

今天，个人比以往任何时候都更加活跃于城市之中。人们通过日常活动就可以为城市的构成添砖加瓦，他们不仅会消费他们所能获得的越来越多的信息，而且更重要的是，还会产生新的数据集。每一个个人移动设备（手

Cities are changing significantly as the consequence of the growing presence of ubiquitous computing and interconnected mobile devices. Increasingly, architecture is made of the complex combination of physical elements (for example walls, roof and facades) and immaterial parts (data, Internet connections, radio signals, GPS tracking and many other wireless technologies).

Perhaps more extensively than buildings, cities are embodying data in their infrastructures. With city dashboards we can, for example, have an overview of how our cities are performing almost in real time. At any moment, citizens can know the level of pollution in a certain area of the city, the exact position of taxis and buses, or have information about particular public or private events.

Moreover, urban data analytics allow for wider and deeper considerations about the performance of cities. This includes the level of wealth of certain neighborhoods, the index of safety in specific areas, or the overall quality of public transport. Citizens are today more aware of how the city in which they live works and the way in which it changes and evolves over time. People can now see aspects of their urban life that were hidden or difficult to explain before the digital age. The digital infrastructure that characterises our cities today provides people not only with a new gaze on the events that are happening around them but, more interestingly, with new insights on their own public and urban lives.

More than ever before, individuals are an active part of the city. Through their daily actions, people contribute to the making of cities not only by consuming the increasing amount of information available to them but, more importantly, by producing new datasets. Every personal mobile device (phones, tablets and the like), activity tracker (wearable technology), audio/video recording device

机、平板电脑等)、活动跟踪器(可携带技术)、音频/视频记录设备和每一次社交媒体互动都有助于创造与真实生活日益交织在一起的数字公共生活。通过社交媒体分享某一事件的图片成为融入公共生活的一个组成部分,就像在拥挤的公共广场上举行真实会议一样。

  建筑师和规划师将这种新的复杂事物融入他们的工作中,采用新的科学方法来进行城市设计和建筑设计。建筑学包括今天新产生的一些学科门类,如城市分析学、感知城市学、城市信息学、响应式景观、互动式建筑等,其中空间概念与编码、脚本和编程直接相关。现如今,建筑通过改变形状、调整边界和使其自身变得更具渗透性来嵌入新的场景模式中。随着我们意识到城市的每个部分像每个人的所有行动一样,都可以被监控、记录和存储在一些基于云的远程数据库中,公共空间的性质也在发生着转变。空间是可以测量的,距离是很容易计算的,只要有信息,我们就(几乎)可以(并可能)实时地了解一个地方的一切情况。

  从历史的角度来看,我们可以认为自己正处于有利的位置,因为我们正在见证建筑环境的巨大变化,这些变化正以惊人的速度发生。每年都会有一项新发明、一项新专利或一项新技术问世。我们的城市将在10~20年后大不相同。作为建筑师和设计师,我们的工作就是接受这些变化,通过建筑环境的设计来改善人们的生活。作为市民来说,我们正处于最好的情况之中,可以轻易地享受到在我们眼前发生的这种转变。

and social media interaction contribute to the making of the digital public life that is increasingly intertwined with the physical one. Sharing pictures of an event through social media is an integral part of the public life as much as the physical meeting into a crowded public square.
Architects and planners are incorporating this new complexity into their work, embracing new scientific approaches to urban and building design. Architecture today includes new strands like urban analytics, sentient cities, urban informatics, responsive landscapes, interactive architecture, etc., where the notion of space is directly connected to coding, scripting and programming. Architecture is now embedding this new scenario by changing its shapes, adapting its boundaries and becoming more permeable. The nature of public spaces is shifting as a consequence of the awareness that every part of the city, like any action of each individual, can be monitored, recorded and stored in some remote cloud-based database. Places are measurable, distances are easily calculated and, as long as information is available, we can know (possibly) everything about a place in (almost) real-time.
Looking through a historical perspective, we can consider ourselves in a favourable position, as we are witnessing these great changes in the built environment that are happening at a galloping speed. Every year there is a new invention, a new patent filed or a new technology available. Our cities will be significantly different in 10-20-year time. As architects and designers, it is our job to embrace these changes and improve people's lives through the design of the built environment. As citizens, we are in the best position to simply enjoy the happening of this transformation before our very eyes.

Norwegian Scenic Routes

挪威的风景线

2. Havøysund
1. Varanger
3. Senja
4. Andøya
5. Lofoten
6. Helgelandskysten
7. Atlanterhavsvegen
8. Geiranger-Trollstigen
9. Gamle Strynefjellsvegen
10. Rondane
13. Gaularfjellet
11. Sognefjellet
14. Aurlandsfjellet
12. Valdresflye
15. Hardanger
16. Hardangervidda
17. Ryfylke
18. Jæren

1. Varanger > Steilneset minnested

Steilneset Memorial, 2012, Peter Zumthor and Louise Bourgeois

2. Havøysund

Selvika, 2012, Reiulf Ramstad Architects

3. Senja

Bergsbotn, 2008, CODE: Architecture AS

4. Andøya

Kleivodden, 2013, Landskapsfabrikken AS - Inge Dahlman

5. Lofoten

Rassikring Vest-Lofoten, 2014, Knut Hjeltnes AS Sivilarkitekter MNAL

Eggum, 2007, Snøhetta

6. Helgelandskysten

Ureddplassen, 2018, Haugen/Zohar Arkitekter

7. Atlanterhavsvegen

Eldhusøya, 2014, Ghilardi + Hellsten Arkitekter AS

8. Geiranger - Trollstigen

Trollstigen, 2012, Reiulf Ramstad Arkitekter AS

8. Geiranger - Trollstigen

Gudbrandsjuvet, 2010, Jensen & Skodvin Arkitektkontor

9. Gamle Strynefjellsvegen

Videfossen, 1997, Jensen & Skodvin Arkitektkontor

10. Rondane

Sohlbergplassen, 2006, Carl-Viggo Hølmebakk

Strømbu, 2008, Carl-Viggo Hølmebakk

11. Sognefjellet

Vegaskjelet, 1996, Carl-Viggo Hølmebakk

15. Hardanger

Steinsdalsfossen, 2014, Jarmund/Vigsnæs AS Arkitekter MNAL

13. Gaularfjellet

Utsikten, 2016, CODE: Architecture AS

17. Ryfylke

Allmannajuvet, 2016, Atelier Peter Zumthor

挪威的风景线
Norwegian Scenic Routes

Per Ritzler

挪威一个主要的现代旅游景点被称为国家风景线,这是一条汇聚了挪威全国各地18条精选公路的风景线,这些公路都穿行在美丽变幻的风景中。

风景通道显然不是挪威所特有的现象,但挪威项目的独特之处在于它有意识地使用了具有现代感的,通常是壮观的建筑。所有的18条路线的休息区、观景点、卫生间设施、通道、楼梯和家具都经过专门设计,大部分是由挪威建筑师设计的,并且越来越受到世界的关注和认可。建在"自然中的建筑"这一理念作为挪威公路项目的一部分,已经成为一个国际概念。

它始于20世纪90年代中期,这个时期,挪威正在失去吸引国际游客的竞争力。那个时候,挪威政府要求挪威公共道路管理局(NPRA)研究如何让挪威美丽、野生的自然环境与广泛的道路交通网络形成互动。该试点项目选择了四条路线,这四条线路包括沿海岸线、穿过山脉和沿挪威南部峡湾的路线。

几位挪威建筑师受命设计休息区、观景台、卫生间设施和家具设施,他们的设计造型大胆,表达形式也很有魄力。这些设计被放置在试点路线上。在索涅埃勒特,设计师甚至引入了一件以石雕形式出现的艺术品。公路旅行者、政府当局和政治家们都喜欢展现在他们面前的设计,并对该项目信心十足。因此,1998年挪威议会批准了进一步开发永久性国家风景线的计划。

挪威公共道路管理局负责开发国家风景线,将其打造成为一个景点。日常工作就交由挪威公共道路管理局下属的风景线路处完成,该机构的主要工作地点是利勒哈默尔。为了确保旅游线路管理的高质量,挪威公共道路管理局设立了一个外部专家委员会,负责管理旅游线路,并就旅游线路专业方面的事宜向风景线路处提供建议。

建筑委员会的职责是确保沿途风景观景点和休息区有高质量的视觉效果,委员会成员包括一位建筑师、一位景观建筑师和一位视觉艺术家。此外,还任命了一名艺术馆长,以确保具有国际价值的艺术作品成为国家风景线体验的一部分。在20世纪90年代初,设计想法是将项目

One of Norway's major modern tourist attractions is known as the National Scenic Routes – 18 selected roads in various parts of the country, which navigate through beautiful and diverse landscapes.

Scenic byways are obviously not a phenomenon specific to Norway, but what makes the Norwegian project quite unique is the conscious use of modern and often spectacular architecture. Rest areas, viewpoints, toilet facilities, pathways, stairs and furniture along all 18 routes have been specially designed, largely by Norwegian architects, and are receiving increasing attention and recognition far beyond national borders. The idea of "Architecture in Nature" as part of the Norwegian road projects has become an international concept.

It started in the mid-1990s; Norway was losing the competition for international tourists. The Norwegian government at that time asked the Norwegian Public Roads Administration (NPRA) to look into ways of exploiting the interaction between Norway's beautiful, wild nature and the country's extensive road network. Four routes were selected in the pilot project – along the coast, across the mountains, and along the fjords in southern Norway.

A handful of Norwegian architects were commissioned to design rest areas, viewing platforms, toilet facilities and furniture with bold forms and daring expressions. The installations were placed along the pilot routes and, at Sognefjellet, even an artwork was introduced in the form of a stone sculpture. Road travelers as well as authorities and politicians liked what they saw, gained confidence in the project, and in 1998 the Storting – the Norwegian parliament – gave the go-ahead for further development of permanent National Scenic Routes.

It is the NPRA that is responsible for developing the National Scenic Routes into an attraction, and the day-to-day work is done in the NPRA Scenic Routes Section, whose main office is at Lillehammer. In order to ensure high quality in the management of Scenic Routes, the NPRA has attached an external expert council that holds the superordinate responsibility and advises the Scenic Routes Section on matters of professional guidelines for their efforts.

An Architecture Council ensures high visual quality of scenic viewpoints and rest areas along the routes, and its members include an architect, a landscape architect, and a visual artist. An art curator has also been appointed to ensure that internationally valuable works of art are included as a part of the National Scenic Routes experience. In the early 1990s

委托给挪威年轻艺术家，让他们参与到风景线项目的建设中来，以此来激发和激励他们。但是，时光荏苒，建筑师也会变老，所以今天与我们合作的是第三代的风景线设计师。此次合作伴随着项目的开发，将一直延续到2023年。到那个时候，国家风景线项目历经30年，将最终完成其目标，在18条线路上将拥有245个建筑设施。

在这里，让我们更仔细地看一下已经安装到位的150个建筑设施中的四个。

挪威南部的哈丹格尔被称为"旅游的摇篮"，这个地方就像一张巨大的明信片，游客可以在这里饱览盛开的果园、宏伟的峡湾和壮观的瀑布。哈丹格尔是瀑布之乡，最壮观的瀑布之一就是位于诺海姆松村庄附近的斯坦达尔斯弗森。瀑布开始激荡流下的地方由不寻常的岩石群构成，游客可以在这里，在咆哮的洪流后面安全地行走，不会弄湿衣服。

斯坦达尔斯弗森在整个旅游季节接待了许多游客，但停车场容量太小，通往瀑布的老路又不平稳，需要维修。因此，该区域被挪威公共道路管理局风景线路处选中，并由Jarmund/Vigsnæs AS Arkitekter MNAL对其进行升级改造（24页）。如今，它拥有一个宽敞的停车场和一座新的服务大楼，内含卫生间设施和旅游信息查询中心。此外，通往瀑布的通道也得到了升级和安全加固。服务大楼通过许多不寻常的角度，限定了人们看向瀑布景观的视野。建筑物和朝向主干道的墙壁是用混凝土建造的，用模板现场浇筑，颜色为特殊的绿色，与周围美丽的环境相致。这条小路被设计成一条"驴道"，台阶很浅，可以轻松惬意地走近瀑布。新斯坦达尔斯弗森景区于2014年开门迎客。

去高拉埃莱的路从沿着峡湾的美丽的小村庄巴勒斯特兰开始，经过一段距离之后公路会急剧蜿蜒上升到山的最高点。从新的休息区Utsikten——"景色"——可以清楚地看到1500m高的山脉和深深的峡谷。Utsikten休息区位于海拔700m处，可以将其描述为一个大型三角形混凝土板，其转角呈向上弯曲的姿态（32页）。

挪威公共道路管理局的风景线路处举办了一次设计邀请赛，对国家风景线高拉埃莱的沿线观景点进行升级改造，2008年该竞赛的桂

the idea was already to stimulate and encourage young Norwegian artists by commissioning them to contribute to the Scenic Routes project. But architects too, will grow old, and so today we are working with the third generation of Scenic Routes designers. These will be following the developments toward 2023, when the National Scenic Routes project reaches its goal after 30 years, with 245 architectural installations along the 18 routes.
In this feature, let us take a closer look at four of the 150 architectural installations that are already in place.
Hardanger in southern Norway is referred to as the "cradle of tourism": the place is like one big postcard where visitors may feast their eyes on blossoming orchards, majestic fjords and magnificent waterfalls. Hardanger is the land of waterfalls, and one of the most spectacular examples is Steinsdalsfossen, near the small village of Norheimsund. An unusual rock formation at the point where the waterfall starts its main drop makes it possible for visitors to walk – safe and dry – behind the roaring torrent.
Steindalsfossen receives many visitors throughout the tourist season, but the car park capacity was too limited and the old path to the waterfall was unstable and in need of repair. The area was therefore selected by the NPRA Scenic Routes Section and upgraded by Jarmund/Vigsnæs AS Arkitekter MNAL (p.24). It has a spacious car park and a new service building with toilet facilities and tourist information. In addition, the pathway up to the waterfall has been upgraded and secured. The service building with its many untraditional angles frames the view towards the waterfall. The building and the wall towards the main road are built in concrete, poured in place against formwork panels, in a special green color that harmonizes with its beautiful surroundings. The path is designed as a "donkey path" with shallow steps that make the walk up to the waterfall nice and easy. The new Steindalsfossen attraction opened in 2014.
The drive to Gaularfjellet starts from the beautiful little village of Balestrand along the fjord before the road begins its sharply meandering climb up to the highest point of the mountain. From the new rest area Utsikten – "the View" – there is a clear view towards 1500m-high mountains and deep canyons. Utsikten, itself at 700m above sea level, can be described as a large triangular concrete slab with corners that are bent upwards. (p.32)
The NPRA Scenic Routes Section invited design entries to an architecture competition for the upgrading of the view-

冠被CODE: Architecture AS摘得。建筑师们的雄心壮志是在兼顾统一性和多样性的基础上实现整体的休息体验。通过建立大量的不同空间和通向翘起的转角的通道，建筑师实现了这一雄心壮志，在这些翘起的转角处可以欣赏到不同的景色。平台的几何形状在建筑结构和自然环境之间创造了一个明显的边界，赋予建筑物一种强烈而独立的表达方式，而这种表达方式结合了现场环境的特点，可以说是此处场景所独有的。混凝土板的使用是为了产生壮观的效果，但同时，它又是一种简单易懂的材料。

除了这些观景点外，混凝土板上还有一个圆形剧场和卫生间设施，以及汽车和公共汽车的停车设施。游客可以在混凝土板周围自由地行走，在某些地方甚至可以在混凝土板下方行走。建造Utsikten非常具有挑战性。项目开始的时候，施工现场没有电力供应，整个施工期间只能使用柴油集料。而且，该地区没有互联网接入，手机信号接收不良。此外，山顶的天气总是变幻莫测，这一点非常具有挑战性。同时，混凝土、钢筋等材料的运输路线较长，在冬天的时候，上山道路还会封闭。

挪威北部的埃尔德胡斯亚岛是"大西洋旅游公路"上的一站，"大西洋旅游公路"是世界上最美丽的海岸公路之一。由于靠近著名的、以惊艳的风景而闻名的斯托塞森德桥，埃尔德胡斯亚岛在1989年路线刚开通的时候就受到来自世界各地的游客的欢迎。埃尔德胡斯亚岛服务区（12页）的设计灵感来源于与地形弯曲率相匹配的道路或铁路的形状，并使用了海洋挖掘所采用的材料和组装方法。对1~17m的斜坡的测量显示了三维曲线的不同变化，同时也说明了该地形的局限性和可能性。长廊下方是咖啡馆、旅游信息查询中心和全年开放的卫生间，为游客提供各种便利。150m长的正立面被用作挡土墙，以释放地面上的空间，方便机动车辆的停放和行驶。立面上的窗户和装饰物是鲱鱼鱼群的抽象艺术表现形式，这主要是参照了艾弗里的捕鱼传统而设计出来的。

本项目所采用的材料和技术方案的选择均可以保证在没有后期管理的情况下至少能使用50年。道路结构组件是用316型不锈钢配件提前预制的，长度进行了缩减，可自由调节，并用螺栓固定其形状。

point along National Scenic Route Gaularfjellet, which was won by CODE: Architecture AS in 2008. The ambition of the architects was to embrace the entire rest experience with both unity and diversity. This was achieved by establishing a number of different spaces and ways to move through turned-up corners addressing different landscape situations. The geometrical shape of the platform creates a distinct boundary between the structure and its natural surroundings and gives the architecture a strong and independent expression which at the same time is totally site-specific. The concrete slab is meant to be spectacular but at the same time simple and instantly readable.
In addition to the viewpoints, the concrete slab has an amphitheater and toilet facilities, as well as parking facilities for cars and buses. Visitors may walk in the terrain, around the concrete slab and, in some places, even underneath it.
Building Utsikten was quite challenging. There was no electricity at the construction site, and a diesel aggregate was used throughout the construction period. There was no Internet access in the area, and cell phone reception was poor. Additionally, the weather at the top of the mountain was always unpredictable and challenging. The transportation route for concrete, reinforcement and other materials was long, and the road up to the mountain was closed during the winter.
Northern Norway's Eldhusøya Island is a stop on the journey of the "Atlantic Tourist Road", one of the most beautiful coastal roads in the world. Because of its proximity to the famous Storseisundet Bridge, which is renowned for its stunning scenery, Eldhusøya Island became popular with tourists from all over the world as soon as it opened its route in 1989.
The Eldhusøya service area (p.12) was inspired by the shape of the road or railway that matched the terrain curvature, and utilized materials and assembly methods from marine excavators. Measuring the slope from 1 to 17m it shows the various changes of the three-dimensional curve, and it confirms the limitations and possibilities of the terrain. Below the promenade are cafés, tourist information centers, and toilets that are open throughout the year to provide convenience for visitors. The 150m long facade performs as retaining wall in order to liberate flat ground for vehicular parking and maneuverings. The windows and decorations on the facade are an abstract representation of a shoal of herrings, referencing Averøy's fishing tradition.

沿着挪威西南部的国家风景线莱菲尔克的旅程是一次充满了对比的旅程，其中包括曝露的高地和郁郁葱葱的绿色山丘对比，陡峭的山坡和深邃的峡湾对比。撒乌达菲雷有着独特的粗糙、原始和顽固的天性。在这里，我们还发现了阿尔曼纳朱韦特锌矿以及在19世纪晚期关闭的矿藏。

位于阿尔曼纳朱韦特的锌矿于1882年投入运营，直至1898年，共生产了12 000t的矿石。这些矿石通过汽船沿撒乌达奥尔登河运输并销往海外。在早期，矿石是经过包装从矿山运输到峡湾的。骡子可以运100kg的矿石，工人每人可以运25kg。采矿作业是索达小镇后来水电站和工业发展的先驱，是该镇文化历史的重要组成部分。

2002年，挪威公共道路管理局风景路线处委托瑞士建筑师彼得·卒姆托在阿尔曼纳朱韦特设计了一个景观来欢迎游客的到来，同时也使得古老的采矿历史焕然一新（46页）。卒姆托在进行建筑设计时受到了采矿作业、苦工和工人们艰苦的日常生活的启发。

为了确保这个历史遗址不会失去它独特的自然和文化品质以及作为一个矿业区域的历史，建筑师选择了一种低调的表达方式。他设计了一系列建筑，包括一座展示采矿历史的博物馆建筑、一家咖啡馆和一座位于主干道旁的服务建筑。处在阿尔曼纳朱韦特峡谷中的这些简单而壮观的建筑是在工业传统的基础上设计的，并与周围的景观融为一体。

建筑师彼得·卒姆托在阿尔曼纳朱韦特建造了壮观的建筑。他再次在挪威为那些面临可怕的困难和残酷的命运的人们树立了一座历史纪念碑。2011年夏天，女王陛下在瓦尔德开设了斯泰尔内塞特纪念馆，这是由挪威公共道路管理局委托彼得·卒姆托建造的第一座设施建筑。阿尔曼纳朱维特是为纪念采矿作业和工人在19世纪晚期的艰苦生活而出现的，而坐落在瓦尔德的斯泰尔内塞特纪念馆是为纪念在17世纪的芬兰被巫术迫害的受害者而建立的。

All the materials and technical solutions in this project are chosen to last for at least 50 years without post-management. The structural components of the path are prefabricated with Type 316 stainless steel fittings, with reduced length, which are freely adjustable, and bolted to fix the shape.
The journey along National Scenic Route Ryfylke in south-western Norway is a journey through contrasts along naked highlands, lush green hills, precipitous mountainsides and deep fjords. Saudafjellet has a distinctive rough, raw and recalcitrant nature. Here we also find Allmannajuvet and the closed-down ore mines dating back to the late 19th Century.
The zinc mines in Allmannajuvet were in operation from 1882 to 1898 and produced 12,000 tons of ore. The ore was transported by steamboats along Saudafjorden and sold overseas. In the early years, the ore was transported by packsaddle from the mines to the fjord. The mules carried 100 kilos of ore; the men each carried 25 kilos. The mining operation was a precursor to the later development of hydropower plants and industries in the small town of Sauda, which is an important part of the town's cultural history.
In 2002, the NPRA Scenic Routes Section commissioned Swiss architect Peter Zumthor to design an installation at Allmannajuvet for the purpose of welcoming visitors and bringing the old mining history back to life (p.46). Zumthor's buildings are inspired by the mining operation, the drudgery and the workers' strenuous everyday lives.
In order to ensure that this historical site would not lose its unique natural and cultural qualities and the experience of a mining community, the architect chose a low key expression. The family of structures include a museum building for presentation of the mining history, a café and a service building by the main road. The simple yet spectacular buildings in the Allmannajuvet gorge are designed within an industrial tradition and fit well into the landscape.
With his spectacular buildings in Allmannajuvet, architect Peter Zumthor has once more erected a historical monument in Norway to the people who once faced terrible hardship and cruel fates. In the summer of 2011, Her Majesty the Queen opened Steilneset Memorial in Vardø, the first of Zumthor's installations commissioned by the NPRA. While Allmannajuvet emerges as a monument to the mining operation and the workers' life of hardship in the late 19th Century, Steilneset in Vardø was raised in memory of the victims of the government's witchcraft persecutions in Finnmark in the 17th Century.

埃尔德胡斯亚
Eldhusøya
Ghilardi+Hellsten Arkitekter AS

位于挪威西海岸的大西洋旅游公路，是今天斯堪的纳维亚旅游人数最多的休闲旅游目的地之一。由于邻近壮观的斯托塞森德桥，该地区国际旅游业的热度持续增加，因此，在1989年7月公路开通后不久，埃尔德胡斯亚这条路很快就开始出现交通拥堵的问题。

该项目由Ghilardi＋Hellsten建筑事务所设计，一条高架步行道、一个停车场和一座小型服务建筑被整合到一个基础设施中，使道路旅行者在岛上能够畅通无阻地步行，同时提供全景的视野以及可供休息和沉思的休息区。

700m平台及其栏杆的悬空特性可以防止游客与开阔景观之间发生直接接触，而具有渗透性的平台能够允许风和雨水的通过，起到滋养下面的当地植被的作用，同时可以保持道路干燥，以确保全年运输的畅通。

在设计概念中，埃尔德胡斯亚公路充分利用了此处的交通基础设施，考虑了其弯曲的模式如何能够适应周围的自然景观，并向海上和近海制造业借鉴材料和预制技术。建筑师根据地形的可能性和局限性进行了复杂的三维曲线变化设计，从而确定了道路的几何形状，并根据地形进行了严格的调整，最大的坡度比达到了1:17。

游客服务设施被嵌在行人基础设施的下方，这里向人们提供了一处避风港，还设有一间咖啡馆、信息查询点和卫生间。这些设施即使在淡季也会向游客开放。150m长的立面被用作挡土墙，以释放出地面空间用于机动车辆的停放和行驶。墙面开窗和装饰的设计灵感来源于艾弗里的捕鱼传统，借鉴了鲱鱼鱼群的抽象性概念。

由于该地区地理位置偏远，气候条件恶劣，而且没有淡季旺季之分，全年的使用都很频繁，因此项目使用的所有材料和技术解决方案都是以免维护作为标准的，预期寿命可达50年。该道路的结构组件是用316型不锈钢预制的，是一套可调节的零件套件，它是一套伸缩构件系统，通过可调螺栓组装并固定在一起。这项技术可以在变化的地形上构建具有毫米精度的完美多面曲面，避免现场焊接以及后来的腐蚀风险。

平台板材由常用于石油钻井的预制纤维复合板条箱构成，它是模块化系统，与下面的结构相匹配，所有的横向连接都好像是在现场切割的一样吻合。立面覆层也由轻质复合板制成，通过CNC铣削并形成

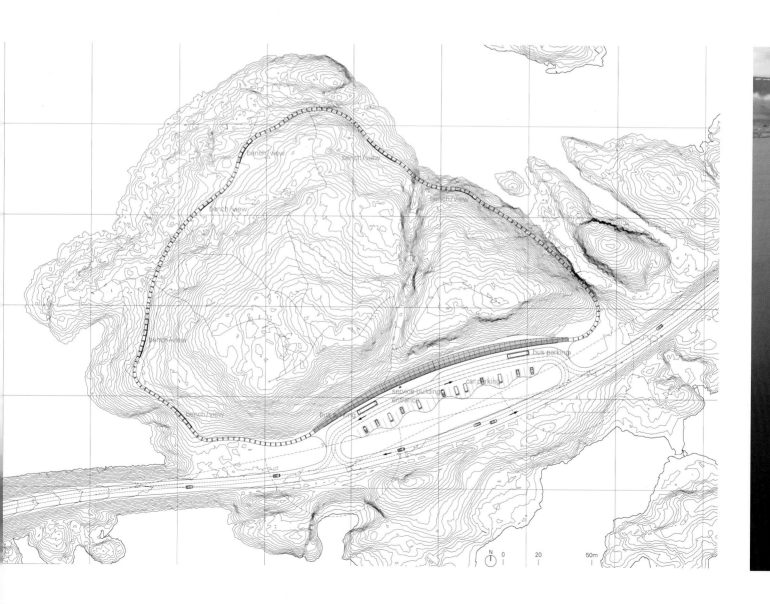

精确的小窗洞及鱼形装饰。栏杆也由不锈钢制成;扶手处的几何结构复制了平台和景观之间的高度,从而形成了一个与下方的地形相呼应的重建的镜像方案。

    本项目的设计和工程的建造旨在保护自然环境,同时为各个年龄段或不同身体条件的游客提供无限制的步行通道,它与标志性的斯托塞森德桥一起,意在强调自然和人造工程之间的关系。

The Atlantic Tourist Road, on the west coast of Norway, is one of the most visited leisure-travel destinations in Scandinavia today. Because of its proximity to the spectacular Storseisundet Bridge, Edhusøya quickly began to suffer traffic congestion shortly after the road opening in July 1989, due to an increase in international tourism in the area. The project, by Ghilardi+Hellsten Arkitekter, integrates an elevated promenade, a car park and a small service building into a single infrastructural facility, offering the road travelers a smooth pedestrian break around the island while providing panoramic views and sitting areas for resting and contemplation.

The hovering nature of the 700m platform, and its railing, discourages direct contact between visitors and the open landscape, while the permeable deck allows wind and rainwater to filter through, nurturing the local flora below, whilst keeping the path dry for transit all year round.

In its conception, Eldhusøya draws from transport infrastructure and how its curving patterns adapt to the natural landscape, borrowing materials and prefabrication techniques from maritime and offshore manufacturing. The ge-

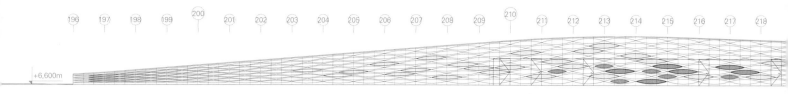

南立面——服务建筑
south elevation _ service building

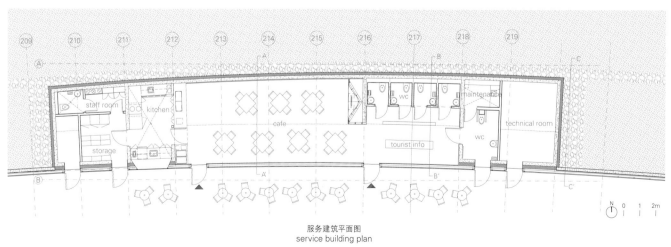

服务建筑平面图
service building plan

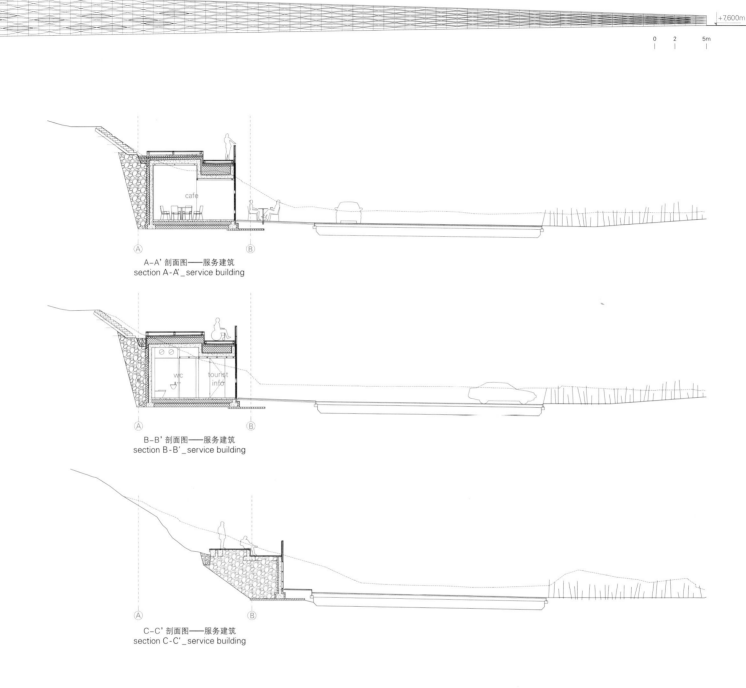

A-A' 剖面图——服务建筑
section A-A'_service building

B-B' 剖面图——服务建筑
section B-B'_service building

C-C' 剖面图——服务建筑
section C-C'_service building

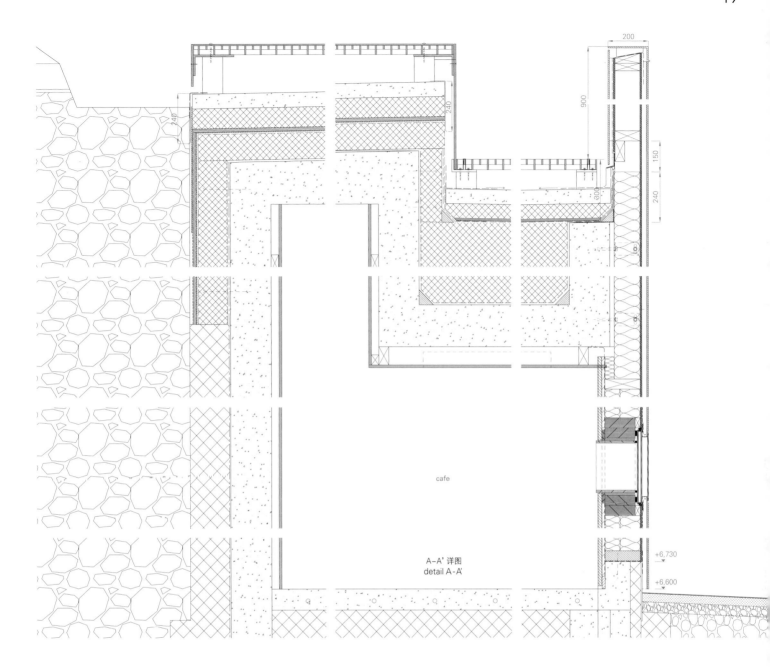

A-A' 详图
detail A-A'

项目名称：Eldhusøya / 地点：Eldhusøya, 6530, Averøy, Norway / 事务所：Ghilardi + Hellsten Arkitekter AS / 项目团队：Franco Ghilardi, Erik Stenman, Anna Nilsson / 室内设计：Ghilardi + Hellsten Arkitekter AS / 景观设计：Ghilardi + Hellsten Arkitekter AS in collaboration with Asplan Viak AS / Thea K. Hartmann (MNLA), Ragnhild Nessa (MNLA) / 工程师：Structural - K. Apeland; Electrical - Norconsult AS; Hvac - UC VVS og Klima Rådgivning AS; Fire - BST / 咖啡馆运营：Quirien van Oirschot
项目管理：Erik Stenman - Ghilardi + Hellsten Arkitekter AS / 施工管理：Sveinung Myklebust - Norwegian Public Roads Administration / 承包商：Betonmast Røsand AS
生产商：Facade - Steni AS; Steel - Trondheim Stål AS; Komposite - Stangeland Glassfiber Produkter AS / 客户：National Tourist Routes, Norwegian Public Roads Administration
市政当局：Averøy / 建筑面积：viewing platform - 1,890m²; service building - 150m² / 竣工时间：2014.7 / 摄影师：©Roland Halbe - p.18, p.19; ©Roger Ellingsen (courtesy of National Scenic Routes) - p.12~13, p.17ˡᵉᶠᵗ, p.21ᵘᵖᵖᵉʳ, p.23; ©Jan Andresen (courtesy of National Scenic Routes) - p.17ʳⁱᵍʰᵗ, p.21ˡᵒʷᵉʳ; ©Jiri Havran (courtesy of National Scenic Routes) - p.16; ©Kjetil Rolseth (courtesy of National Scenic Routes) - p.15; ©Rolf-Ørjan Høgseth (courtesy of the architect) - p.20

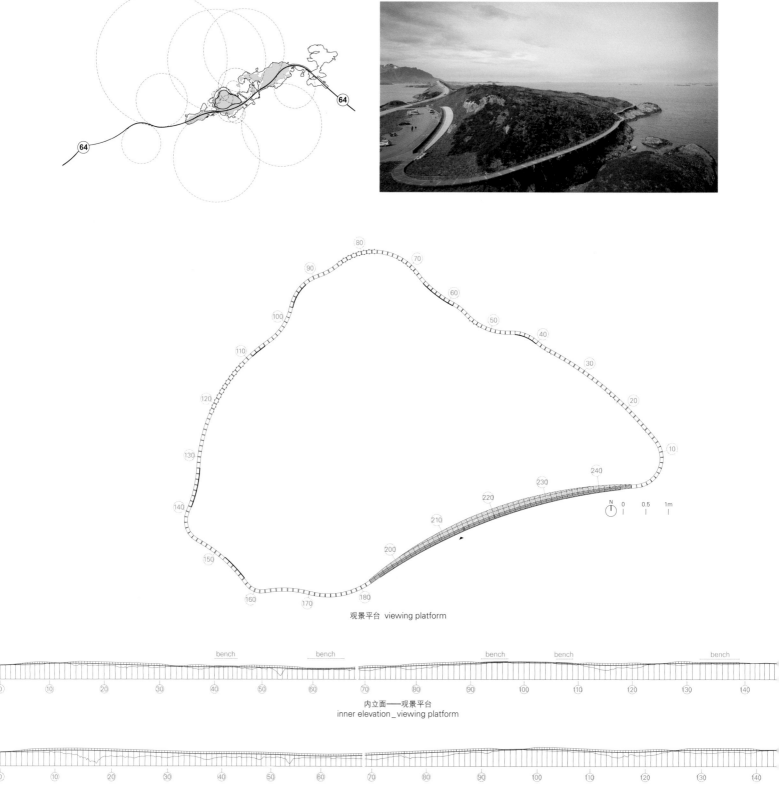

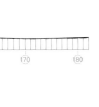
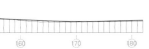

ometry of the path reacts to the possibilities and limitations of the terrain through a complex variation of tridimensional curves. These are tightly tailored to the topography, with a maximum slope ratio of 1:17.

Tourist services are embedded under the pedestrian infrastructure providing a weather refuge with café, information-point and restrooms, which are also open out of season. The 150m-long facade acts as a retaining wall in order to liberate flat ground for vehicular parking and maneuvering. The wall fenestrations and the décor are inspired by Averøy's fishing tradition featuring an abstract shoal of herring.

Given its remote location, rough weather and intense year-round use, all the materials and technical solutions used in this project are maintenance free, with a nominal life expectancy of 50 years. The path's structural components are prefabricated in Type 316 stainless steel as an adjustable kit-of-parts, a system of telescopic members, which are assembled and fixed together in place by means of adjustable bolts. This technique allows constructing perfect faceted curves with millimetric precision over the variating topography, avoiding on-site welding and its subsequent risk of corrosion.

The platform deck is made of prefabricated fiber composite crates commonly used on oil rigs, which are modular systems and which match the structure below with all its transversal joints "cut to fit" on site. The facade cladding is also made of lightweight composite panels that were CNC milled to produce the precise eye-window holes and its fish-pattern décor. The railing is also made of stainless steel; the geometry of the handrail copies the heights between platform and landscape all along the path, resulting in a reconstructed mirrored profile of the terrain just below.

This project is designed and engineered to protect the natural environment while providing unlimited pedestrian access to all kinds of visitors, regardless age or physical condition. Together with the iconic Storseisundet Bridge, this intervention attempts to underscore the relationship between the natural and manmade.

观景平台剖面图
viewing platform sections

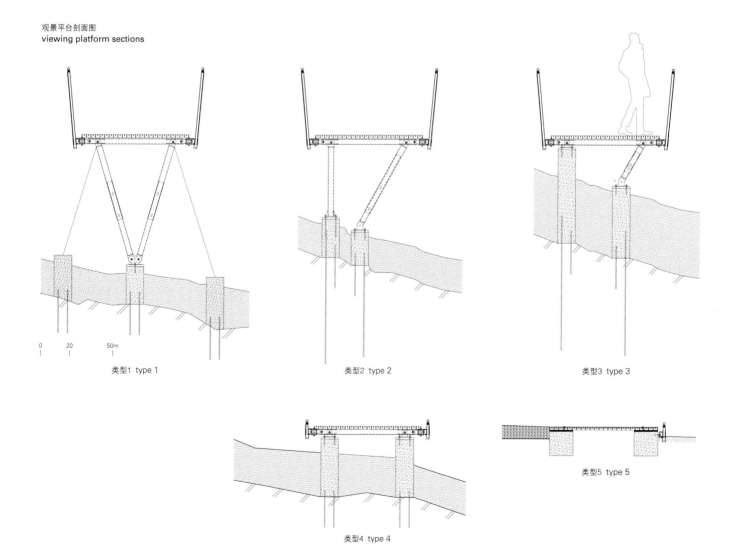

类型1 type 1

类型2 type 2

类型3 type 3

类型4 type 4

类型5 type 5

挪威议会和挪威政府已经授权挪威公共道路管理局开发国家风景线，包括18条途经挪威美丽变幻景观的精选道路。其目的是通过大胆的建筑，提升对自然风光的体验，使挪威成为一个更具吸引力的旅游目的地。

Jarmund/Vigsnæs建筑师事务所的建筑师们被委托在哈丹格尔的斯坦达尔斯弗森进行该项目的设计开发工作。一百多年以来，该地区那郁郁葱葱的绿色植被、瀑布、山脉和冰川一直吸引着游客的到来。

该项目包括一个停车场、一个旅游信息查询中心、一个卫生间设施以及一条通往主要景点——斯坦达尔斯弗森瀑布的人行道。停车场和旅游设施所处的场地位于瀑布对面，中间隔着一条河，并沿着一条废弃道路的曲线设置。该项目旨在将项目本身融入景观之中，同时将停车场和旅游设施与瀑布联系起来。该项目能将游客的目光引向瀑布，同时还能阻隔来自道路的视线。

建筑师设计了倾斜的混凝土墙来调节道路和停车场之间的高度差。这面墙已成为项目的一条脊椎，它沿着道路，呈曲线形状，充当着整座建筑的脊梁。

卫生间和旅游信息查询建筑为具有雕塑感的混凝土体量，这些混凝土体量通过有顶户外空间与倾斜墙体连在一起。混凝土体量逐渐远离墙体，扭曲着指向瀑布的方向，同时也反映出道路的运动轨迹以及瀑布的静态视图。

在内部，这种扭曲使得身处旅游信息查询中心室内空间的人们可以通过一个大窗户欣赏到被框定住的瀑布景观。从外部看，这种扭曲在停车场和青草坡道之间形成了一个入口，这条坡道能引导游客前往瀑布。

混凝土是该项目的主要材料。现浇混凝土使具有雕塑感的建筑体量可以实现墙体和屋顶之间没有差异，从而使整个项目拥有统一的形式。混凝土中添加了一种二氧化铬颜料，表面呈绿色。这种绿色将建筑与景观联系起来。此处的景观主要由绿绿的田野和植被组成，另外，在春天，河流和瀑布也呈现出绚丽的绿。混凝土模子使用的是粗糙的木板，这样做就使墙体和屋顶朝外的一侧呈现出粗糙但又均匀的纹理。但是，面向室内的一侧是光滑的。

休息区周围的行人区和旅游信息查询中心，以及通往瀑布和瀑布

斯坦达尔斯弗森瀑布
Steinsdalsfossen Waterfall

Jarmund/Vigsnæs AS Arkitekter MNAL

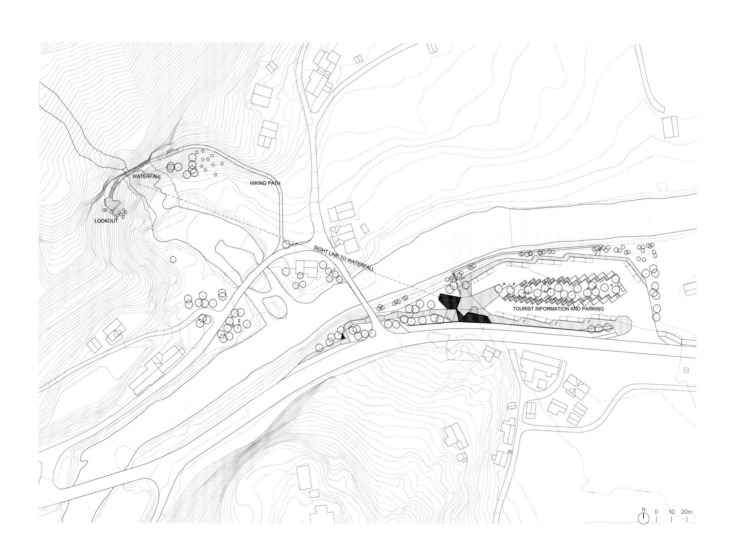

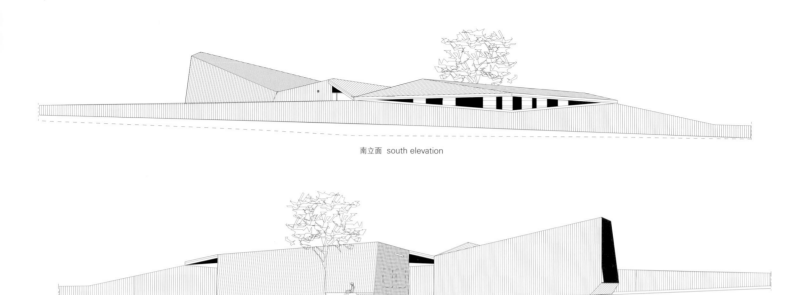

南立面 south elevation

北立面 north elevation

后面的人行道，都是使用经过刷子处理的灰色混凝土制成的，外形呈棱角状。灰色混凝土使用耐候钢条作为框架，同时也利用耐候钢条进行细分。

　　该项目提供给人们的不仅是对现实道路旅行生活的突破，更是一种非凡的自然体验，在其中，建筑给人们提供了欣赏景观的全新方式。

The Norwegian Public Roads Administration has been commissioned by the Norwegian Parliament and Government to develop the National Scenic Routes, consisting of 18 selected stretches of road passing through Norway's beautiful and varied landscape. The objective is to turn Norway into a more attractive visitor destination, by enhancing the experience of the magnificent natural scenery with bold architecture.

Jarmund/Vigsnæs Architects was commissioned to develop the destination at Steinsdalsfossen, Hardanger. Lush green vegetation, waterfalls, mountains and glaciers in the area have attracted tourists for more than a hundred years. The project consists of a parking area with a tourist information center and a restroom facility, in addition to a footpath leading to the main attraction – Steinsdalsfossen waterfall. The site for the parking area and tourist facilities is situated across the river from the waterfall, along the curve of a trafficked road. The project aims to integrate itself into the landscape, while connecting the parking and tourist facilities to the waterfall. The project simultaneously directs views towards the waterfall as well as providing a barrier from the road.

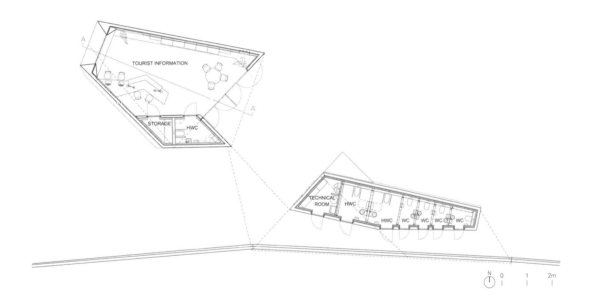

A slanted concrete wall is introduced to mediate the difference in levels between the road and the parking area. This wall has become the spine of the project, following the curve of the road and acting as the backbone of the buildings.

The washrooms and tourist information buildings are sculpted concrete volumes physically connected to the wall through covered outdoor spaces. As the volumes gradually separate away from the wall, twisting and pointing towards the waterfall, they simultaneously reflect the movement of the road as well as the static view of the waterfall. Internally, this twist allows for the interior of the tourist information center to be focused on a framed view of the waterfall through a large window. Externally, the twist creates a portal between the parking area and the green grass ramp that leads the tourists towards the waterfall.

Concrete is the main material of the project. In situ concrete allows the sculpted building volumes to be constructed in a way that does not differentiate between walls and the roofs, giving the project a cohesive form. Added to the concrete is a chromium-dioxide pigment, resulting in a green surface. This green color links the building to the landscape, which is dominated by green fields and vegetation as well as the almost fluorescent green color of the river and waterfall in the springtime. The concrete molds are of coarse planks, giving the walls and roofs of the volumes a rough and uniform texture on the outside. However, the ceiling and wall surfaces facing covered areas are smooth.

The pedestrian zones around the rest area and tourist information, as well as the footpath leading to and behind the waterfall, are made of brushed gray concrete in angular shapes. The gray concrete is framed and subdivided by strips of Corten steel.

The project delivers not only a practical break from life on the road, but also a dramatic experience of nature where architecture enables new ways of seeing the landscape.

项目名称：Steinsdalsfossen Waterfall / 地点：Hardanger, Norway / 事务所：Jarmund/Vigsnæs AS Architects MNAL
项目团队：Einar Jarmund, Håkon Vigsnæs, Alessandra Kosberg, Siv Hofsøy, Claes Cho Heske Ekornaas, Rakel Helling, Jens Herman Næss, Martin Blum-Jansen, Ane Groven / 客户：Statens Vegvesen, Turistvegprosjektet / 顾问：Grindaker AS, Multiconsult As Bergen, Lysstoff AS
用途：tourist information + service facilities / 建筑面积：110m² (tourist information and restrooms ca) / 施工时间：2008—2014
摄影师：©Jiri Havran (courtesy of the architect) - p.25[upper], p.27, p.29[upper]; ©Nils Petter Dale (courtesy of the architect) - p.25[lower], p.26, p.28; ©Roger Ellingsen (courtesy of Norwegian Scenic Routes) - p.29[lower], p.31

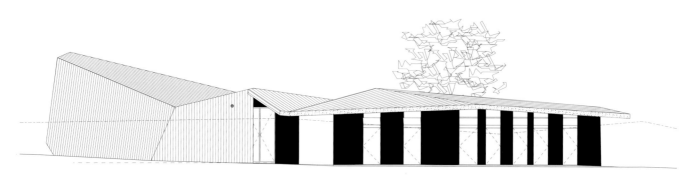

南立面——旅游信息查询中心+卫生间
south elevation _ tourist information + toilet

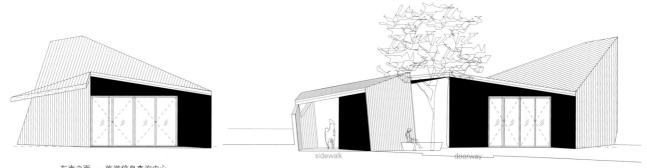

东南立面——旅游信息查询中心
south-east elevation _ tourist information

北立面——旅游信息查询中心
north elevation _ tourist information

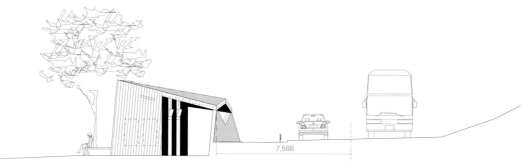

西立面——卫生间
west elevation _ toilet

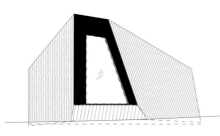

西北立面——旅游信息查询中心
north-west elevation _ tourist information

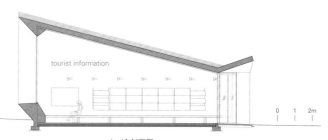

A-A' 剖面图
section A-A'

Utsikten 休息区
Utsikten
CODE: Architecture AS

高拉山，休憩佳处

2016年，CODE: architecture AS建筑事务所在一个名为Utsikten（景色）的小场地完成了本案项目。它位于绵延的高拉山脉中，是挪威西海岸之路上一个天然的停车眺望点。经过长途跋涉，徒步穿越群山的人们来到这里，都会被眼前壮丽的起伏山脉所震撼。而对于驾驶着车辆一路向北的人们来说，这里则是回望他们所路过一系列惊险的急转弯的最佳场所。

CODE建筑事务所在Utsikten设计的项目是一个巨大的三角混凝土平台，就位于路边。平台的三个尖角如同羽翼般向外向上扬起，让80cm厚的平台仿佛轻盈地落在地面上。如同野餐的台布被整洁地布置，创造出恰如其分的环境来迎接一场盛宴一样，Utsikten的混凝土平台也被设计成一个可供人们全方位驻足的佳所。车辆可以一直开到观景台边上，而游客们则可以在尖角之间穿行，从不同的角度，全方位地欣赏壮美的群山景色。

为了确保完工项目拥有最佳的位置，并能够充分地利用场地的不同特质，事务所的建筑师们在场地中搭建出全尺寸的概念模型。借助吊车和绳索，建筑师设计了电线模型结构，经过几轮的完善，才形成3D模型、图纸，进而建成最终的建筑。

混凝土结构的处理工作极具挑战性，这是由于山脉位置特殊，结构几何形状复杂，而且对结构质量要求高。为了尽可能地做到万无一失，参与项目的各方都首先参加了一个结构测试，这个结构已经建造了平台的一大部分，这样做的目的是确保各方对最终的结构达成共识，并对此前提出的选择和方法进行质量方面的评估。

完工后的平台呈现独立的几何结构特征，耸立在延绵的群山风景之中。材料与建造技艺的选择低调粗犷，而造型却又独树一帜。随着时间的流逝，混凝土将会风干，植被将会慢慢覆盖在平台之上，平台的颜色也会不断变化，恍如自然山体的一部分。

由粗钢管制成的扶手顺着平台的边缘而立，通透的铁网在保证安全的同时丝毫没有遮挡视线。混凝土的表面经过了碾磨、抛光、喷砂或封装，建筑的形态和分区也因而更为明确。高耸的角落留出了可供站立、小坐的位置，还有通往卫生间和山坡的门洞，以及供雨水流出的泄水口。而在雨天，这些飞扬的角落又成为可供人们临时容身、躲避风雨的场所。

The Gaular Mountain – Set for Rest

In 2016 CODE: architecture AS completed the project at the site known as Utsikten (The View), a natural stopping

place when driving over the Gaular mountain area along the western coast of Norway. For those who have made the trek over the mountain itself, the site affords a surprising vista overlooking a majestic landscape. Conversely, for those driving up, Utsikten constitutes the ever-visible apex of the ascent rising up through a series of hairpin bends.
CODE's project at Utsikten is a large, triangular concrete platform situated right at the side of the road. The platform is 80cm thick but appears to rest lightly atop the terrain with raised, wing-like corners that protrude outwards and upwards into the air. Similar to how a picnic blanket is neatly arranged in order to create the proper setting for a meal, the concrete platform at Utsikten has been developed to create a proper setting for an entire stopover. The cars may drive all the way up to the outlook, and visitors can move around between the corners to experience the spectacular scenery from various angles.

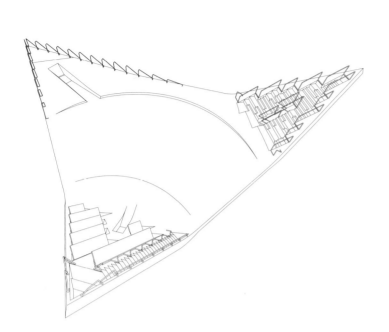

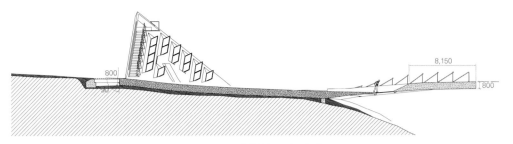

A-A' 剖面图 section A-A'

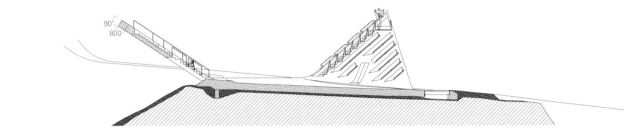

B-B' 剖面图 section B-B'

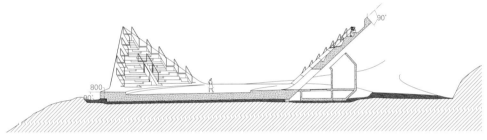

C-C' 剖面图 section C-C'

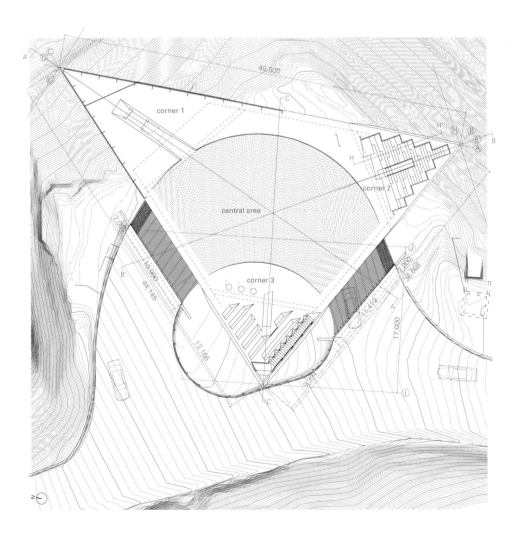

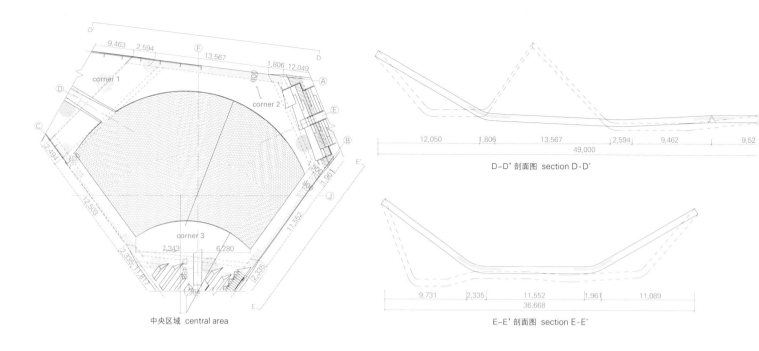

中央区域 central area

D-D' 剖面图 section D-D'

E-E' 剖面图 section E-E'

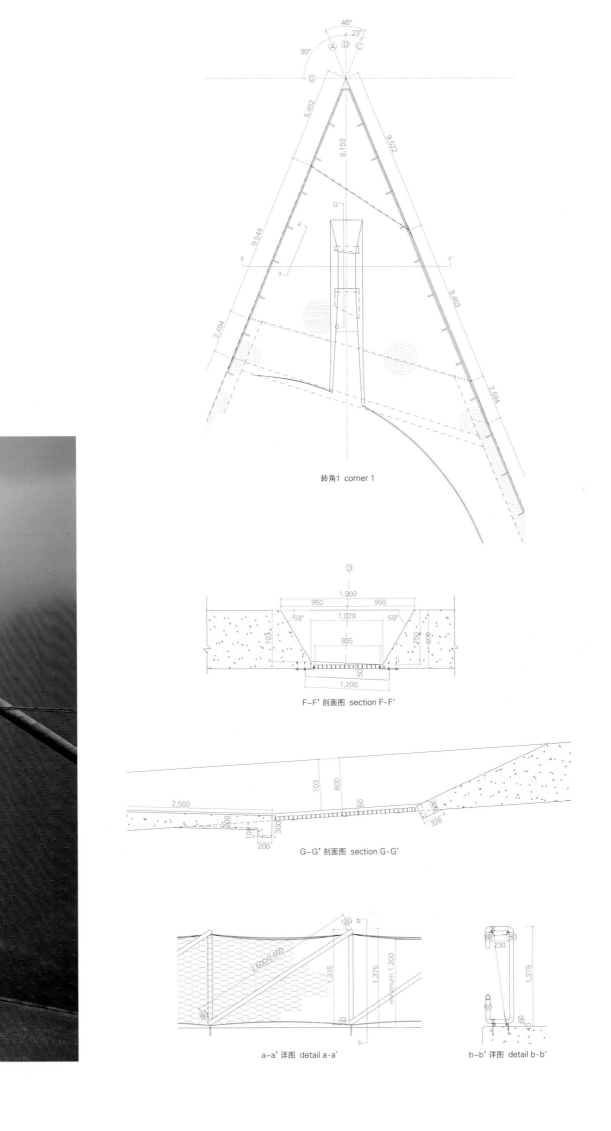

转角1 corner 1

F-F' 剖面图 section F-F'

G-G' 剖面图 section G-G'

a-a' 详图 detail a-a'

b-b' 详图 detail b-b'

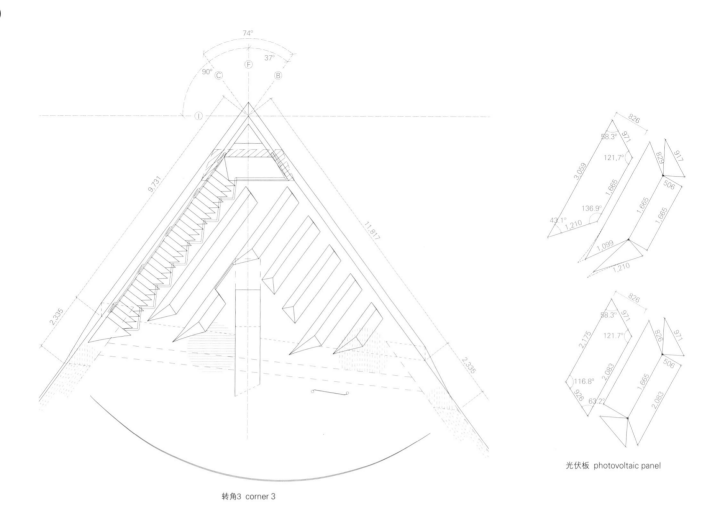

转角3 corner 3

光伏板 photovoltaic panel

项目名称：Utsikten
地点：Gaularfjell, Balestrand municipality, Western part of Norway
事务所：CODE: architecture AS
工程师：B-consult by Steinar Bjercke and DIFK by Florian Kosche
开发商：National Tourist Routes
承包商：Veidekke Sandane
面积：850m² / 竣工时间：2016.6
摄影师：
©Jiri Havran (courtesy of the architect) - p.35 bottom, p.36 bottom, p.43;
©Jarle Wæhler (courtesy of Norwegian Scenic Routes) - p.34, p.35 top, p.35 middle, p.38~39, p.40, p.42; ©Per Ritzler/Statens vegvesen (courtesy of Norwegian Scenic Routes) - p.32~33, p.41; ©Trine Kanter Zerwekh/Statens vegvesen (courtesy of Norwegian Scenic Routes) - p.44~45

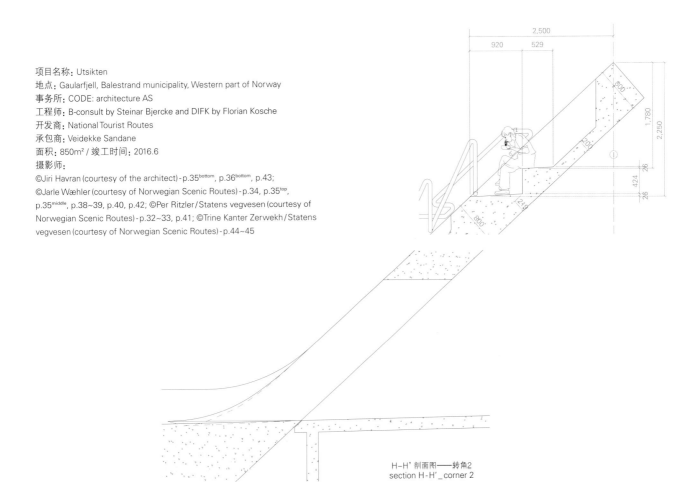

H-H'剖面图——转角2
section H-H'_corner 2

In order to ensure the optimal placement for the finished project and exploit the site's various qualities, the architects sketched the concept at full scale on the site itself. With the aid of a crane and ropes, the structure was constructed as a wire model in several rounds before finally being translated into a 3D model, drawings, and the finished product.

The work on the concrete construction proved to be quite challenging because of the mountain's location, the complex geometry, and the high demands of quality. In order to be as prepared as possible, the different parties involved in the construction project first participated in a trial construction of large sections of the platform in order to participate in common experiences and to evaluate quality assurance of the given choices and approaches.

The finished platform appears as an independent, geometric, and precise object in the landscape. The materials and technology are familiar and robust, while it is the shape itself that is spectacular. Over time the concrete will weather, acquiring natural vegetation, and its color will begin to approximate that of the surrounding mountainsides.

The rails have been made from thick steel pipes as a visually conspicuous seam along the platform's edges and reliefs, with outstretched nets that are more or less transparent. The concrete includes surfaces that have been milled, honed, sandblasted or board sheathed in order to accentuate the platform's shape and zonal divisions. The raised corners contain holes and reliefs that provide sitting and standing room, access to the restrooms, and egresses to the terrain, or that serve as an outlet for the rainwater amassed at the platform. During rainy weather it is also possible to seek shelter beneath these corners.

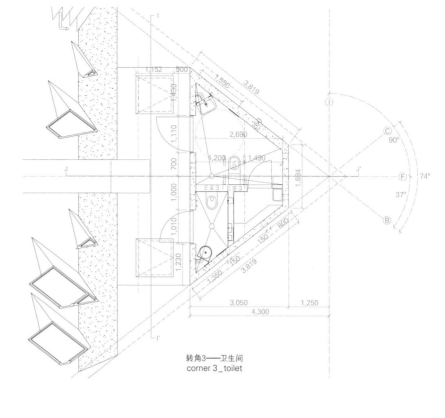

转角3——卫生间
corner 3_toilet

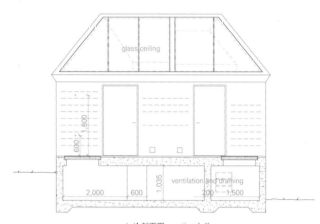

I-I'剖面图 section I-I'

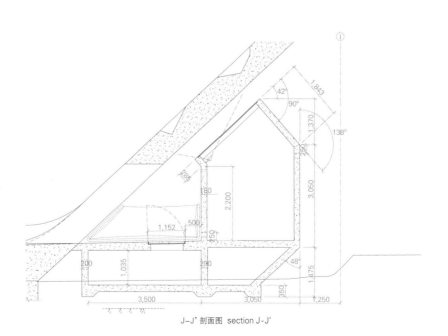

J-J'剖面图 section J-J'

阿尔曼纳朱韦特锌矿博物馆
Allmannajuvet Zinc Mine Museum
Atelier Peter Zumthor

1882年，挪威索达的阿尔马纳峡谷开启了锌矿开采的历史。在一条出峡谷的蜿蜒小径上，骡子把一车车的矿石从矿井拖到了悬崖的边缘，在这里，矿石被抛到山谷下，并被粉碎成小块。这些碎块被清洗并运到大约10km之外的索达港，然后运到英格兰进行加工。该矿于1899年关闭，主要是由于锌在全球的市场价格发生了变化。

在一块外露的岩石上，矿石早已被冲走。这里曾经坐落着一个矿工营房，现在是520国道上的一个休息站，隶属于挪威旅游公路网。这些公路蜿蜒1800km，覆盖全国，从南到北，好像不断在向游客们招手，招呼他们来这个特别的景点驻足一下，欣赏美丽风景并倾听那段历史故事。

瑞士建筑师彼得·卒姆托受挪威公共道路管理局的委托，为这座被遗忘的矿山注入新的活力。经过仔细观察，人们确实可以发现峡谷采矿历史的痕迹：从矿山入口的地方开始的运输路径经过切割与斜坡支撑墙和桥梁融为一体，从前用于抛掷矿石的木质平台的基础，还有早已消失的简单木结构的基础遗迹。

设计工作就是要与这些构件打交道。博物馆由四个结构组成，它们是沿旧矿道设置的轻型木结构。朴实的露天博物馆在重新设计的休息处开始它的第一站。在这里可以看到结构组的第一个成员——服务大楼。在休息处的对面是一段新的石阶，从凹坑处通向以前的小径。再往前走几步就是矿区咖啡馆，这在很大程度上是为夏季的游客服务的，但也可以全年被索达和周边地区的居民全年用来举办小型活动。这里提供简单的当地食品和饮料，以及当地人制作的衣服和家居用品。建筑师对"旧方法，新形式"的调查是该设计最重要的前提，重要的是让索达当地人参与到设施的使用和运营中来。

在峡谷通道再转个弯之后，人们就来到了一个避风港和一个集合点，在那里，在导游带领下来参观的人会得到头盔和灯等设施。在隔壁，游客们可以爬上矿业博物馆，这是建筑群的第四位也是最后一位成员，以前，正是在这里，矿石被从悬崖上抛下去。

在长期从事收集信息工作的当地居民莱维·阿里德·贝尔德的帮助下，建筑师们收集了博物馆可用的所有文件。在这个小博物馆的展品中，展出的东西很少，但仍然令人印象深刻：库存证明、购买合同、保险文件、时间表、旧照片和几件采矿设备。

矿井里的工作很辛苦，但证明这一点的文件似乎消失得很快。因此，事务所委托历史学家阿恩维德·利勒哈默尔写了一本关于该矿历史方面的书籍，而地质学家斯坦·埃里克·劳瑞岑则绘制了该矿及其地理特征的地图。来自于索达当地的作家基雅尔坦·福莱格斯塔德编撰了一本关于地下主题的世界文学的选集：《Sub Terra – Sub Sole》。这三本书，每一本都很独特，且都由Aud Gloppen设计，并在博物馆里展出。

Zinc mining in the Almanna Canyon in Sauda, Norway, commenced in 1882. On a trail snaking out of the canyon, mules dragged cartloads of ore from the mine to the edge of the cliff, where it was hurled down to the valley floor, shattering into smaller pieces. These fragments were washed and transported about ten kilometers to the harbor at Sauda to be shipped for processing in England. The mine was closed in 1899; the world market price for zinc had changed.
On a rocky outcrop where the ore was washed and a miners' barrack once stood, there is now a rest stop on National Highway 520, belonging to the Norwegian Tourist Highway Network. These roads run 1800km through the country from south to north, repeatedly beckoning visitors to stop at special sites to enjoy the beauty of the landscape and its history.
Swiss architect, Peter Zumthor, was commissioned by the Statens vegvesen, the Norwegian Public Roads Administration, to lend new life to the forgotten mine. On close inspection, one can indeed discover traces of the canyon's mining history: the transport trail beginning at the mine en-

摘自专著《彼得·卒姆托1985—2013年。建筑与项目》。文字、草图、图纸和水彩画/彼得·卒姆托，编辑/托马斯·杜里奇，Scheidegger&Spiess出版社，2014年（ISBN 978-3-85881-723-5）

Extract from the monograph *Peter Zumthor 1985-2013. Buildings and Projects*. Texts, sketches, drawings and watercolors by Peter Zumthor, edited by Thomas Durisch, Scheidegger & Spiess, 2014. (ISBN 978-3-85881-723-5)

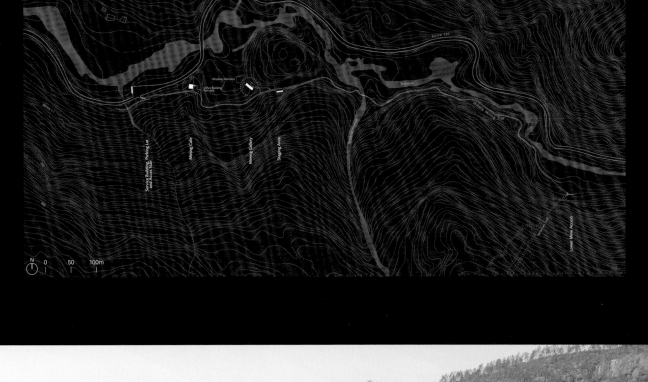

trance cut into the slope with supporting walls and bridges, the foundations of the wooden platform from which ore was thrown, and the remains of foundations for simple wooden structures long since disappeared.

The design works with these elements. It comprises a family of four structures – light wooden constructions along the old mine pathway. The modest open-air museum begins at the redesigned rest stop. Here one finds the first family-member, a service building. Opposite the rest stop is a new flight of stone stairs leading to the former trail from the pit. A few more steps up is the mining cafe. This largely serves tourists in summer but can be used for small events all year round by the residents of Sauda and environs. Here, simple local foods and beverages are available as well as clothing and household items made by locals. An overriding preoccupation for the architects was the investigation of "old methods – new forms"; it was important to involve the people of Sauda in the use and operation of the facility.

After a further turn in the canyon pathway one comes upon a shelter, a collection point where participants on a guided tour are given helmets and lamps. Next door, visitors climb up to the mining museum, the fourth and last member of the family of buildings, perching on the cliff where ore was once thrown down.

With the help of Leiv-Arild Berg, a local resident who had long been gathering information, the architects assembled all available documents for the museum. The pickings were slim – the exhibit in the little museum shows this – but nonetheless impressive: stock certificates, purchase contracts, insurance documents, timesheets, old photographs,

and a few pieces of mining equipment.

Work in the mine was backbreaking, but documents which prove this seem to disappear quickly. So the practice commissioned historian Arnvid Lillehammer to write a history of the mine, and the geologist Stein-Erik Lauritzen to draw a map of the mine and its geology. Writer Kjartan Fløgstad, originally from Sauda, put together an anthology of world literature on the subject of being underground: Sub Terra – Sub Sole. These three books, each unique and designed by Aud Gloppen, are on show in the museum.

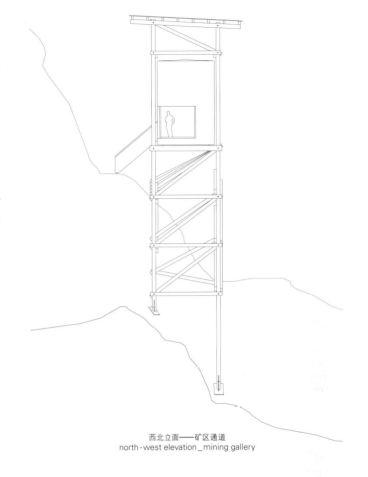

西北立面——矿区通道
north-west elevation _ mining gallery

项目名称：Allmannajuvet Zinc Mine Museum
地点：Allmannajuvet Ravine, Sauda, Norway
事务所：Peter Zumthor
合作者：Atelier Zumthor - Maximilian Putzmann (Project leader), Matthew Bailey, Lisa Barucco, Melissa de la Harpe, Caroline Hammarstroem, Niels Lofteröd, Pavlina Lucas, Simon Mahringer, Sofia Miccichè, Gian Salis, Stephan Schmid, Rainer Weitschies, and Annalisa Zumthor-Cuorad
结构工程师：Finn-Erik Nilsen, Jürg Buchli and Lauber Ingenieure AG
施工监理：Inge Hoftun, Konsul AS, Maximilian Putzmann, Atelier Peter Zumthor & Partner AG
总承包商：Mesta AS, Per-Albert Rasmussen
平面设计：Aud Gloppen, Blaest Design AS
工艺经理：(Statens vegvesen): Arne O. Moen
项目所有者：(Statens vegvesen): Jan Andresen, Harald Bech-Hanssen
客户：Statens Vegvesen, National Scenic Routes in Norway
材料：Scaffolds - creosote pressure impregnated nordic pine gluelam, hot dip galvanised steel parts, corrugated zinc sheets;
Boxes - pressure impregnated nordic pine gluelam, plywood, PMMA resin coating system with Jute inlay, pigmented and linseed oil treated concrete, oil burned steel parts, stained birch;
Furniture - made from black alder wood
规划开始时间：2003
施工（干砌工作与基础）时间：2009—2012
施工（建筑）开始时间：2013
竣工时间：2016.9.8
摄影师：©Aldo Amoretti-p.47, p.51, p.54, p.58, p.59, p.60upper, p.61;
©Per Berntsen-p.50top, left-middle, center, left-bottom, middle-bottom, right-bottom, p.52, p.57upper, lower, p.60lower; ©Fredrik Fløgstad/Statens vegvesen (courtesy of Norwegian Scenic Routes) -p.50right-middle, p.56~57left; ©Jiri Havran/Statens vegvesen (courtesy of Norwegian Scenic Routes)-p.48~49

矿区通道平面图
mining gallery plan

54

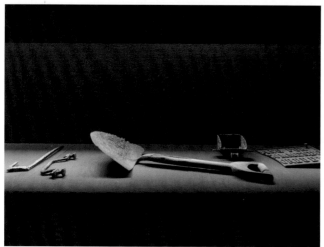
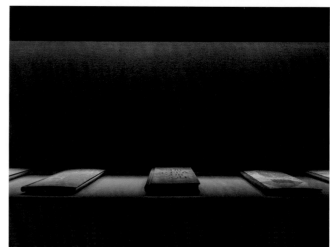

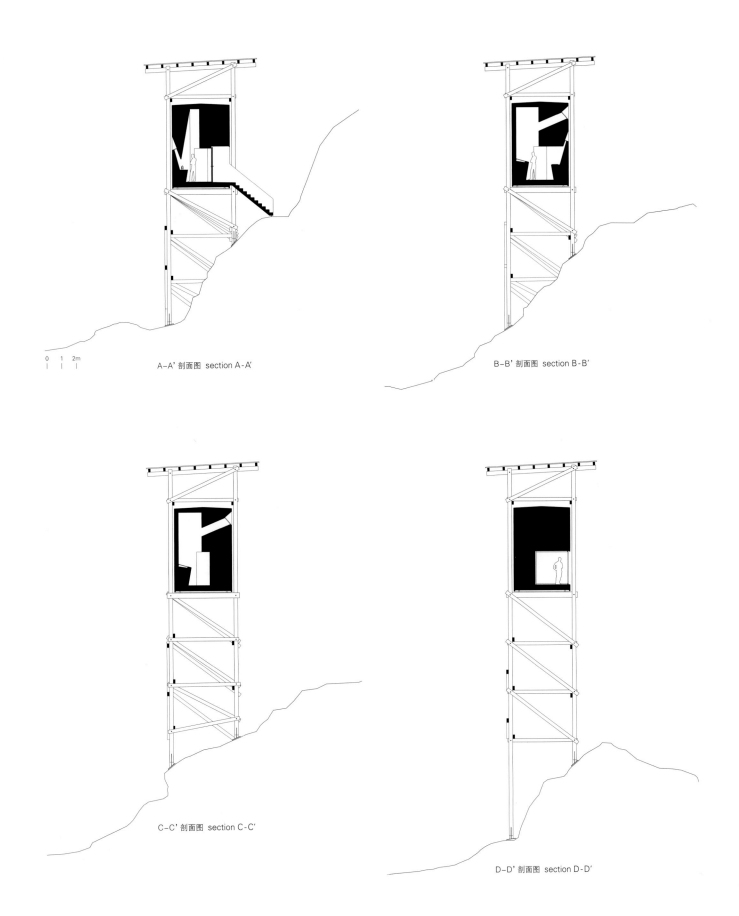

A-A' 剖面图 section A-A'

B-B' 剖面图 section B-B'

C-C' 剖面图 section C-C'

D-D' 剖面图 section D-D'

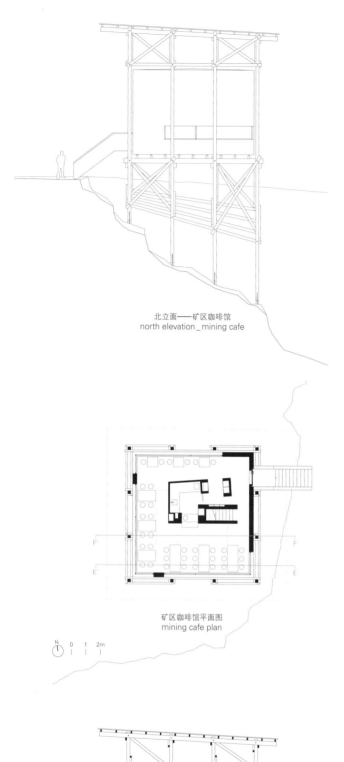

北立面——矿区咖啡馆
north elevation_mining cafe

矿区咖啡馆平面图
mining cafe plan

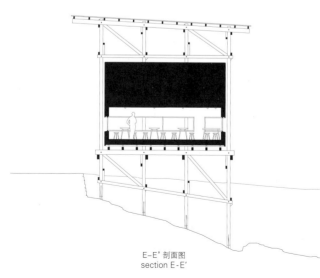

E-E' 剖面图
section E-E'

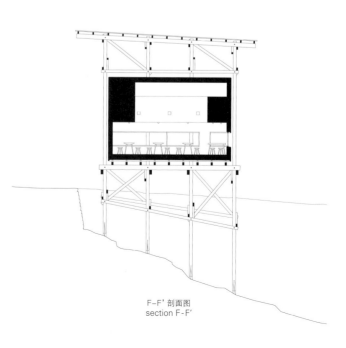

F-F' 剖面图
section F-F'

新旧熔融

Additi`
Integrating

建筑师今天面临的主要任务是重新设计开发现有的建筑结构。特别是在中欧，约三分之二的建筑活动集中在城市地区[1]，重点已经从开发新项目转向住宅和公共建筑的再利用和扩建。无论是通过改变容积、交通方式、外观还是改变其与周围环境的关系，都需要通过加建才能使现有的结构适应不断变化的生活模式。在更深层面上，加建反映了我们与过去的关系：我们所重视的是我们所保留的，以及我们如何

The main task architects are facing today is the redevelopment of the existing built fabric. Especially in Central Europe, where approximately two thirds of building activity is in urban areas,[1] there is a shift of emphasis from new development to the reuse and extension of residential and public buildings. It is through architectural additions that the existing fabric adapts to evolving living patterns, whether through changes in capacity, circulation, external appearance or in relation to the surroundings. On a deeper level, additions reflect our relationship with the past: what we value is reflected in what we keep and how we integrate

布里斯托尔老维克剧院_Bristol Old Vic/Haworth Tompkins
乌托邦——图书馆和表演艺术学院_Utopia – Library and Academy for Performing Arts/KAAN Architecten
Musis Sacrum音乐厅_Musis Sacrum/van Dongen-Koschuch
Coal Drops Yard购物中心_Coal Drops Yard/Heatherwick Studio
红十字会志愿者之家_Red Cross Volunteer House/COBE
普拉托新Pecci当代艺术中心_New Pecci Center for Contemporary Art in Prato/NIO Architecten

新旧熔融_Additions: Integrating Old and New/Isabel Potworowski

将其与新事物结合起来。在这方面,最近的加建项目显示出新旧结合的趋势: 当代建筑元素往往是开放的, 并与公共空间建立新的联系, 与旧建筑形成对比, 同时强调其历史特征。影响新旧建筑融合的历史因素, 揭示了历史建筑的品质以及它们在当代再开发中的潜力。尤其需要注意的是, 历史建筑的价值包括它们提供稳定性、持续性和真实性的能力以及它们对场所营造所做的贡献。

it with the new. In this respect, recent architectural additions reveal a certain trend in the combination of old and new: contemporary building elements are often open and establish new connections with public space, contrasting with the older buildings while emphasizing their historic characters. The historic factors that have influenced the integration of old and new shed light on the qualities of historic buildings, and on their potential for contemporary redevelopment. In particular, the value of historic buildings includes their ability to provide a sense of stability and continuity, their authenticity, and their contribution to placemaking.

新旧熔融
Additions: Integrating Old and New

Isabel Potworowski

稳定性和连续性

哪些结构应该保留以及如何使它们与较新的建筑结合的问题与西欧现代化息息相关。这带来了从永恒的传统社会到进步中的启蒙思想的转变；现代化的理念也提出了应该保留什么的问题。[2] 它塑造了建筑保护运动和我们赋予遗产的价值取向，例如，国家纪念碑的登记和联合国教科文组织世界遗产遗址的认定问题。

很自然，在建筑保护运动出现之前，建筑物就已经被保存了很久。在古代的西方，对旧建筑的关注源于实际的、宗教的和象征性的原因。但是，作为一种意识形态的保护想法只出现在欧洲暴力的政治、社会和经济运动中，特别是1789年爆发的法国大革命。[3] 在之后不久的1790年，第一条保护法被制定了出来。[4] 面对与过去的突然"决裂"，许多人认为需要加入稳定的因素。[5] 从那时起，人们选出了一些历史建筑来作为与现代社会不同的东西。最初的选择仅限于象征性和宗教性建筑，但在20世纪80年代，这种选择扩大到包括工业建筑在内的其他建筑。[6]

定义真实性

如何做出价值判断来决定必须保留什么以及如何保留？历史上，这些问题与真实性有关。随着工业革命的到来，出现了两种对立的建筑运动：一种是提倡真正意义上的新建筑，另一种是寻求保护真正意义上的旧建筑。

Stability and Continuity

The question of which structures to preserve and how to integrate them with newer architecture is tied with the modernization of Western Europe. This brings a shift from the timeless traditional society to Enlightenment ideas of progress; and the very idea of modernization raises the issue of what should be kept.[2] It has shaped the conservation movement and the value that we assign to heritage, for instance with national monument registers and UNESCO World Heritage sites.

Naturally, individual buildings were being preserved long before the emergence of the conservation movement. In Western antiquity, care for old structures stemmed from practical, religious and symbolic reasons. But conservation as an ideology only emerged in the violent political, social and economic movements in Europe, in particular the French Revolution of 1789.[3] Shortly after, in 1790, the first law of preservation was defined.[4] In the face of a sudden break with the past, many felt an urgent need for stabilization.[5] From then on, selected historic buildings were "set apart" as something distinct from modern society. While this selection was at first limited to symbolic and religious buildings, it was broadened in the 1980s to include other structures such as industrial buildings.[6]

Defining Authenticity

How can a value judgement be made to decide what must be preserved, and how? Historically, this question was related to authenticity. With the advent of the Industrial Revolution, two opposing architectural movements arose: one that advocated the authentically new, and the other that sought to protect the authentically old.

Modernism emerged in response to new technologies in construction such as steel, glass and reinforced concrete.

现代主义的出现是对在某些结构中使用新技术的一种回应，这些结构包括钢结构、玻璃结构和钢筋混凝土结构等。这些新技术代表了传统建筑方法的根本性转变，需要一种更适合的建筑方法，这种方法能识别出功能理性主义的新理念，也是现代建筑的真实表现。这一新运动通过全新的方法完全打破了与历史建筑的联系。

约翰·罗斯金曾深深地影响了欧洲19世纪的建筑保护运动。他认为，现代的建筑方式引发了压迫性的功利主义和资本主义的异化。对于他来说，旧建筑的不规则纹理是人类劳动乐趣的至高无上的表现。"一座建筑的最大荣耀不在它的石头上，也不在对黄金的使用上。它的辉煌在于它的时代，在一种深沉的清脆感觉中……这种感觉我们在长久以来被人性的波涛冲刷过的墙壁上可以感受到。"在他看来，历史建筑是"活的纪念碑"，其累积的物质和如画的时代古韵体现了一种真实的理想，甚至是真理性的东西。[7]

在围绕战后纪念碑重建的辩论之后，1964年签署的《威尼斯保护和修复纪念碑和遗址宪章》制定了一套国际性的指南，促进了新旧建筑之间的对比，主张了罗斯金的物质真实性传统。[8]这种尖锐的新旧分离理念让建筑保护主义者和现代建筑师们达成了共识，使建筑保护与战后激进的现代重建得以共存。

这种状况一直延续到20世纪70年代初，当时一种新的保护激进主义试图通过重新引入历史元素来纠正现代主义与历史的突然决裂。这一转变标志着后现代主义新建筑运动的开始，而后现代主义又因其对历史风格表面化的模仿而受到批评。[9]用新的建筑方法建造历史的形式被许多人谴责为"伪造""仿冒"，这些最终都是不真实的。然而，后现代主义对旧建筑的模仿，作为赋予新建筑一种归属感和意义的

These technologies represented a radical shift from traditional building methods, and required a fitting architectural approach that would recognize the new ideals of functional rationalism and that would be an honest expression of modern construction. This new movement was a complete break with historical architecture, embracing a tabula rasa approach.

According to John Ruskin, who greatly influenced the 19th Century conservation movement on the European level, the modern way of building evoked oppressive utilitarianism and the alienation of capitalism. Older buildings' irregular texture was, for him, a supreme expression of the joy of human labor. "The greatest glory of a building is not in its stones, nor in its gold. Its glory is in its Age, and in that deep sense of voicefulness [...] which we feel in walls that have long been washed by the passing waves of humanity." In his view, historic buildings were 'living monuments' whose cumulatively accumulated physical substance and picturesque patina of age embodied an ideal of authenticity, even of truth.[7]

Following the debates surrounding the post-war reconstruction of monuments, the 1964 Venice Charter for the Conservation and Restoration of Monuments and Sites drew up a set of international guidelines that promoted a contrast of new and old, asserting the Ruskinian tradition of material authenticity.[8] This sharp old-new separation found consensus among both conservationists and modern architects, allowing the coexistence of conservation with radical post-war modern redevelopment.

This status quo survived until the early 1970s, when a new conservation radicalism sought to correct modernism's abrupt break with history by reintroducing historical elements. This shift marked the beginning of the new architectural movement of postmodernism, which was in turn criticized for its superficial imitation of historical styles.[9] Building

普拉托新Pecci当代艺术中心，意大利
New Pecci Center for Contemporary Art in Prato, Italy

方式，表达了渴望与历史产生联系的愿望。这一愿望在今天表现为对旧结构物理真实性的重新认识。尤其是，历史建筑的翻修和加建越来越受欢迎，这说明旧建筑变得越来越有品位。"你可以做一个好的当代设计，但你不能构建历史，"建筑师克里斯蒂安·布勒克纳解释说，"现有的多层建筑就像是来自过去的礼物。"[10]

场所营造

在当今全球化的世界里，城市重建面临着没有场地的挑战——想要建造标准化的和不真实的城市环境，却根本没有合适的地方。[11] 在这种情况下，建设有意义的、活跃的公共空间，或是营造场所就变得越来越重要。

创造高质量的公共空间最近在北半球的城市重建开发中显得尤为重要。内城区正在重新开发，以吸引以知识为基础和以服务为基础的经济活动，对于这些活动而言，促进协作的城市环境和紧密的网络至关重要。由于个人收入增加，以及服务行业或知识领域部门的员工更喜欢在内城区居住，因此对城市便利设施的需求也在不断增长。[12] 这些变化使得作为开放会面场所和景点的公共区域的质量变得非常重要。同时，这些变化也导致人们对公共和文化项目多样性的重视程度越来越大。因为城市的再生要介入现有的特定社会和文化历史环境之中，因此保持和加强这段历史的存在感是创造场所感的一个重要方面。历史建筑因其材料的真实性、与之相关的记忆以及在特定历史时期的表现，而给建筑环境增添了一层新的意义。

historic forms with new construction methods was condemned by many as "fake", "pastiche" and ultimately inauthentic. Postmodernism's imitation of the old, however, expressed the desire of connecting with history as a way of giving a sense of rootedness and meaning to new buildings. This desire manifests itself today as a renewed appreciation for the physical authenticity of older structures. In particular, the growing popularity of renovations and additions to historic architecture speaks to a growing taste for the old. "You can make a good contemporary design, but you cannot construct history," explains architect Christian Brückner. "A multi-layered existing building is like a gift from the past."[10]

Placemaking

In today's globalized world, urban redevelopment faces the challenge of placelessness – standardized and inauthentic urban environments without significant places.[11] In this context, building meaningful and lively public spaces, or placemaking, has become increasingly important.

Creating high-quality public spaces has become particularly relevant in recent urban regeneration developments in the Global North. Inner city areas are being redeveloped to attract knowledge-based and service-based economic activities, for which urban environments that foster collaboration and tight networks are essential. There is also a growing demand for urban amenities as a result of increased private income and of service- or knowledge-sector employees who prefer inner city areas.[12] These changes have contributed to an increased importance placed on the quality of public spaces as open meeting places and destinations, and on a diversification of public and cultural programs. Because urban regeneration intervenes in an existing setting with a given social and cultural past, maintaining and strengthening this history is an important aspect of creating a sense of place. Historical buildings add a layer of mean-

乌托邦——图书馆和表演艺术学院，比利时
Utopia – Library and Academy for Performing Arts, Belgium

当代加建

当代建筑的加建体现了历史建筑的价值，通过增加新的功能，使其向周围的公共空间开放并赋予其新的视觉特征来调整历史建筑。新旧之间产生强烈的反差，并突出地勾勒出旧建筑的轮廓，从而以不同的方式重新审视历史建筑的价值。

NIO建筑事务所设计的普拉托Pecci当代艺术中心扩建项目（144页）使原博物馆向周边环境开放，并赋予原博物馆一种新的身份。它是通过投资公共设施和促进社会凝聚力来复兴社区的一项资助计划的一部分，旨在成为新创意区的催化剂。除了将原1988年博物馆的展览空间扩大一倍外，还增加了新的功能，包括档案馆、专业图书馆、礼堂、餐厅和工作坊空间。一个新的公共空间被以大型广场入口的形式添加了进来。与原有的模块化直线建筑不同，加建部分由流线型的圆形建筑组成。一条C形的走廊围绕着原建筑群弯曲状设置，原建筑的一层还建造了相互连接的圆形空间。展览空间与新公共空间产生了强烈的联系，新公共空间的一层非常通透，设有面向展览空间的窗户；一条"天线"从建筑中向上延伸出去，从高速公路上一眼就能辨别出这里是博物馆。

乌托邦（84页）是一个改造项目，KAAN建筑事务所将一座1880年的学校建筑改造成比利时阿尔斯特市的一座图书馆和表演艺术学院，从而创造了新的公共空间，这些空间与旧建筑产生了互动交流。建筑师拆除了原有的庭院式城市建筑的两侧结构，保留了这所具有历史意义的学校，学校的庭院立面增加了一个紧凑的矩形扩建部分，释放了街区的周边空间，从而形成了三个新的公共空间，并连接到具有中世纪韵味的城市街道和广场网络中。在扩建部分内，图书馆被设计成中性的开放空间，与历史建筑的立面形成对比并形成其框架结构。

ing to the built environment because of their material authenticity, the memories associated with them and their representation of a particular historical period.

Contemporary Additions

Contemporary architectural additions reflect the value of historic buildings, adapting them by adding new functions, opening them towards the surrounding public space, and giving them a new visual identity. They contrast and frame the old, reflecting the value of historic buildings in various ways.

NIO Architecten's extension to the Pecci Center for Contemporary Art in Prato (p.144), opens the original museum to the surroundings and gives it a new identity. It is part of a funded initiative to revive the neighborhood by investing in public facilities and promoting social cohesion, and aims to be a catalyst for a new creative district. As well as doubling the exhibition space of the original 1988 museum, new functions are added including an archive, a specialist library, an auditorium, a restaurant and workshop spaces. A new public space is added in the form of a large entrance piazza. Contrasting the original modular rectilinear building, the addition is composed of fluid, round forms. A C-shaped gallery curves around the original complex, and interconnected rounded spaces are built on the ground floor under the original structure. Strong connections are also made with the new public space with a transparent ground floor and windows to the exhibition space; an "antenna" also announces the museum from the highway.

Utopia (p.84), KAAN Architecten's transformation of an 1880 school building into a library and academy for performing arts in the Belgian city of Aalst, creates new public spaces that communicate with the old building. The architects demolish two sides of the existing courtyard-type city block, retaining the historic school. A compact rectangular

Musis Sacrum音乐厅,荷兰
Musis Sacrum, the Netherlands

Coal Drops Yard购物中心,应该
Coal Drops Yard, UK

楼板悬挑伸入这个中心空间之中,在原来的学校和新建筑之间留有空隙。扩建部分的排练室和教学空间采用全高和全宽窗户,与周围的城市空间建立起强有力的视觉联系。

随着1847年Musis Sacrum音乐厅(100页)在荷兰阿纳姆的扩建,van Dongen-Koschuch也开始专注于打造建筑与周围公共空间的联系。扩建部分增加了一个新的大厅和第二个音乐厅,同时向绿色的Musispark公园开放演出场地。与历史上装饰华丽的砖砌建筑形成鲜明对比的是,音乐厅在两块白板之间增加了一个极简的全玻璃单层大厅。从这个"基座"升起的新音乐会场地在舞台后面有一扇大窗户,因此可以看到公园。此窗口也可以为户外活动而打开。

同样,COBE建筑事务所在哥本哈根设计的红十字会志愿者之家(134页)与丹麦红十字会原有总部形成鲜明对比,并向街道开放。最初建于1952年的深灰色建筑立面设计低调内敛,通过一个停车场和一个有围墙的前花园与街道隔开。新增加的设施是为丹麦34 000名红十字会志愿者而打造的,是一个白色的楼梯景观,从三层与老建筑的连接处向下延伸至街道层,创造了一个室外的"城市起居室"。两个小庭院打断了这个阶梯状的屋顶,其中一个通向两栋建筑的新共用入口。在这个新的公共空间下面是两个礼堂和用于培训、会议和活动的多功能室。

Heatherwick工作室将Coal Drops Yard (118页) 改建为购物区,将工业区域与伦敦国王十字广场周围的公共空间重新连接起来。该项

extension is added to the courtyard facade of the school, freeing the perimeter of the block and forming three new public spaces that connect to the medieval city's network of streets and squares. Inside the extension, the library is designed as neutral, open space that contrasts with and frames the historic facade. Floor slabs cantilever into this central void, leaving a gap between the original school and the new building. The extension's rehearsal studios and teaching spaces feature full-height and full-width windows, which establish a strong visual connection with the surrounding urban space.

With the extension to the 1847 Musis Sacrum (p.100) concert hall in Arnhem, the Netherlands, van Dongen-Koschuch have also focused on the connections with the surrounding public space. The extension adds a new lobby and a second concert hall while opening the performance venue to the green Musispark. Contrasting the historic ornate brick building, the addition is a minimalistic, fully glazed single-storey lobby between two white slabs. Rising from this "plinth" is the new concert venue, which features a large window behind the stage that looks onto the park. This window can also be opened for outdoor events.

Likewise, COBE's Red Cross Volunteer House (p.134) in Copenhagen contrasts with the existing headquarters of the Danish Red Cross and opens it towards the street. The original dark gray building of 1952 has an introverted facade and is separated from the street by a parking lot and a walled front garden. The new addition, a facility for Denmark's 34,000 Red Cross volunteers, is a white stair landscape that steps down from its connection on the second storey with the old building to street level, creating an outdoor "urban living room". Two small courtyards puncture this stepped roof, one of them leading to a new shared entrance for the two buildings. Beneath this new public space are two auditoriums, and multipurpose rooms for training, meetings and events.

1. Frank Peter Jäger, Old & New, Basel: Birkhäuser, 2010, p.8
2. Paul Spencer Byard, "Innovation and Insight in the Contemporary Architecture of Additions", Harvard Design Magazine 23, 2005: http://www.harvarddesignmagazine.org/issues/23/innovation-and-insight-in-the-contemporary-architecture-of-additions
3. Miles Glendinning, The Conservation Movement: A History of Architectural Preservation, Abingdon: Routledge, 2013, p.2-3
4. Rem Koolhaas, "Recent Work", GSAPP Transcripts: Preservation is Overtaking Us, 2014: https://www.arch.columbia.edu/books/reader/6-preservation-is-overtaking-us
5. Miles Glendinning, p.2
6. Frank Peter Jäger, p.8
7. Miles Glendinning, pp.117-119
8. Idem, pp.398-400
9. Idem, p.320
10. Frank Peter Jäger, p.11
11. Norsidah Ujang, Khalilah Zakariya, "The Notion of Place, Place Meaning and Identity in Urban Regeneration", Procedia - Social and Behavioral Sciences 170, 2015, p.710
12. Karoline Brombach, Johann Jessen, Stefan Siedentop, Philipp Zakrzewski, "Demographic Patterns of Reurbanisation and Housing in Metropolitan Regions in the US and Germany", Comparative Population Studies 42, 2017, pp.281-284

目避免了创建一个孤立的购物中心,它的设计目标是要建立一条开放式的和具有渗透性的购物街,拥有不同的游览路线和入口。两座维多利亚式铁路建筑分别建于1850年和1860年,用于储存和运输煤炭,它们经过修复的历史立面围合出了新的公共空间。从原有的形式中毫无突兀感地表现出来的新加建建筑是由两座建筑的屋顶形成的,屋顶慢慢向外膨胀,并融合在一起,创造了一个"心脏",遮蔽了下面的聚集空间。

霍沃斯·汤普金斯为一级列管建筑布里斯托尔老维克剧院(70页)所做的加建结构设计是打开建筑的前面部分来容纳更多的观众,并改善剧院与周围公共空间的联系。加建结构取代了1972年的扩建部分,当时的扩建部分有两个不相连的门厅,覆盖了剧院的公共立面,并建有一个过于正式的入口。相比之下,如今这个加建结构的设计为乔治剧院和库珀大厅创造了一个单独的门厅。这一中心空间的焦点是乔治剧院新建的裸露墙体,墙体上铺设了饱经风霜的历史砖块表面,并通过一个采光井照亮。门厅被设计为街道的非正式的延伸部分,由结构木材和玻璃覆盖,创造出明亮、温暖的空间,并框定出历史立面。在街道入口上方,一个红色的霓虹灯标志以"请进"的字样欢迎着游客的到来。

这些当代加建建筑,显示出新旧之间不断变化的关系。今天,这种关系将开放性、公共空间和会议功能的全球价值与历史建筑的独特性和个性品质结合起来。它们的和谐结合创造了一种历史延续感,在拥抱发展的同时,保持着与过去的联系。

Heatherwick Studio's conversion of the Coal Drops Yard (p.118) into a shopping district reconnects the industrial site with the surrounding public spaces of the King's Cross site in London. The project avoids creating an insular shopping mall, aiming instead to design an open and porous shopping street with many routes and entrances. The restored historic facades of the two heritage Victorian rail buildings, which were built in 1850 and 1860 to store and transfer coal, frame the new public space. Seamlessly emerging from the existing forms, the new addition is shaped by the two buildings' roofs that gently bulge outwards, fusing together to create a "heart" that shelters a gathering space underneath.

Haworth Tompkins' new addition to the Grade I listed Bristol Old Vic (p.70) theater opens the front of house areas to a broader audience and improves the connection with the surrounding public space. It replaces a 1972 extension, which had two unconnected foyers, covered the theater's public facade and had an overly-formal entrance. The contemporary addition, in contrast, creates a single foyer for the Georgian theater and Cooper's Hall. Occupying the focal point of this central space is the newly exposed wall of the Georgian theater, its richly weathered historic brick surface illuminated by a light well. The foyer is designed as an informal extension of the street, covered by structural timber and glass that create a bright, warm space and frame the historic facades. Above the street entrance, a red neon sign welcomes visitors with the words "come on in".

These contemporary additions show the evolving relationship between old and new. Today, this relationship combines global values of openness, public spaces and meeting functions with the identity and character-bearing qualities of historic buildings. Their harmonious combination creates a sense of historic continuity, maintaining a connection with the past while embracing development.

布里斯托尔老维克剧院
Bristol Old Vic

Haworth Tompkins

霍沃斯·汤普金斯为英国一级列管建筑布里斯托尔老维克剧院设计完成了一个新的门厅和一座小剧场。

该项目经过了为期五年的精心研究、咨询、设计和施工,旨在将剧院原先正门部分的空间打开,从而吸引数量更多的、兴趣更多元化的观众,同时将剧院置于布里斯托尔公共生活和空间的中心。

建造于1766年的布里斯托尔老维克剧场是目前所有英语母语国家内还在运作的最古老的剧院。时至今日,剧院中的礼堂还大都完好无损。然而,多年连续的改造于1972年达到高潮,于是,当时著名的英国建筑师彼得·莫洛提出,可以设计一个新的门厅和小剧院。2012年,建筑师Andrzej Blonski更新改造了剧院的礼堂和后台。

然而,随着时间的推移,现如今,越来越多的观众脱离了剧院生活。20世纪70年进行的改造,尽管达到了技术层面的要求,但却将礼堂分隔成了两个毫无联系的门厅,使剧院与街道隔绝开来。因此,彻底的改变迫在眉睫。设计团队决定完全拆除和重建1972年加建的门厅和小剧院,取而代之的是更为热情和清晰的门厅空间,从而恢复了库珀大厅内的大公共空间。新建的门厅充当着街道的非正式延伸空间,该门厅由木质框架和玻璃围合而成,以确保阳光能够进入室内空间。

乔治礼堂的设计重点是外立面,现在人们可以从街道上看到礼堂的外立面。立面被一个采光井照亮,其上还设置了新的窗洞,覆盖了原来的老窗户。

霍沃斯·汤普金斯建筑事务所的主管斯蒂夫·汤普金斯这样说道:"改造一座一级列管建筑对我们来说是一个巨大的责任,但是,我希望我们不仅能够将剧院悠久而丰富的独特历史完美地呈现出来,更能够创造一个全新的、生动的公共空间,使每个来这里的人都能有一种宾至如归的感受。"

事实上,门厅中位于夹层空间的画廊、蜿蜒的楼梯和观景平台使得所有观众都能体验这个充满欢乐的独立空间,而且,门厅在白天还同时作为咖啡厅和酒吧使用,为当地社区中的人们提供了一个可以聚会的场所。一个全新的戏剧工作室占据了库珀大厅以前圆筒状的仓库空间。

建筑临街的南立面宛如一件公共艺术品,由可手动调节的遮阳板组成,同时还结合了加里克(18世纪有影响力的演员和剧作家)发表的就职演说和布里斯托尔城市诗人米尔斯·钱伯斯的诗句中的文字。这些文字不仅强调了剧院悠久历史的重要性,也表达了它在布里斯托尔的未来中所发挥的重要作用。

与大部分实际项目中所进行的工作一样,建筑师在选择建筑材料的时候,都会考虑到它们的耐用性和可持续性。因此,建筑师选用花旗松木材作为结构材料,随着时间的推移,它们在颜色上将会逐渐变深,而沿着门厅一侧布置的精致的由橡木板条打造而成的隔板则可以有机地弯曲和伸缩。随着时间的推移和使用进程的推进,铜质的吧台和混凝土地板,以及上层空间中所使用的橡木地板和上了漆的楼梯栏杆也会逐渐变得古色古香。空间中的细节都是非常直观且随意的,使得过渡空间几乎都保留着室外空间的建筑语言,以强调门厅作为街道和历史建筑之间起协调作用的空间的那种感觉。

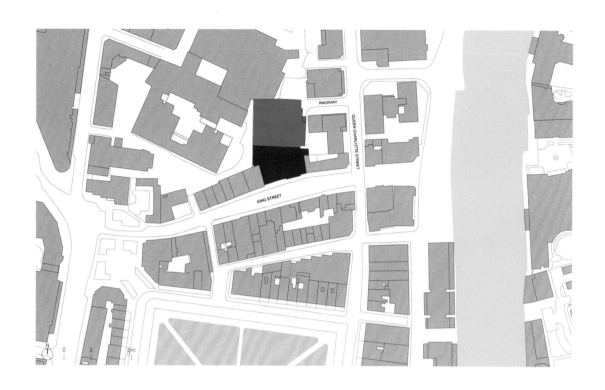

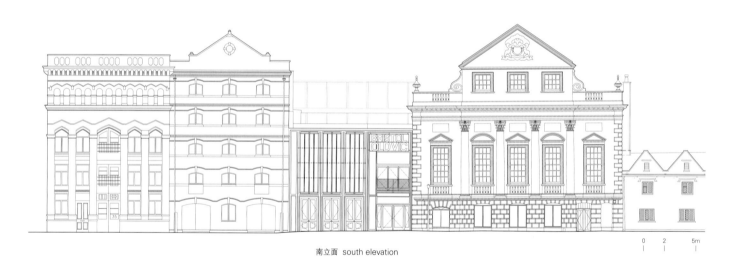

南立面 south elevation

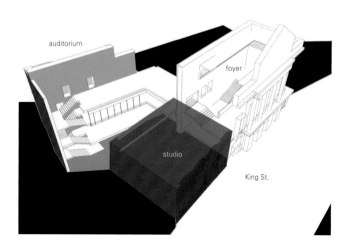

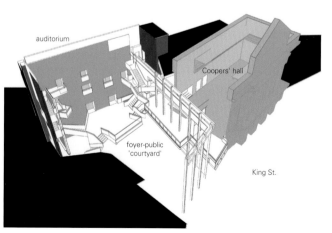

在本项目中，建筑师在结构框架、屋顶和外立面等新建筑构件中使用了结构木材和木衬板。门厅和新建的戏剧工作室通过大型的通风口空间和预冷却室来实现自然通风，同时设置可自动调节温度的高窗和低窗。可移动的遮阳板旨在优化夏季遮阳和冬季取暖的效果。裸露混凝土楼板和整个空间中原有的砖石表面都是良好的蓄热体，同时，拆除作业中的拆下来砖块也被回收再利用，用在新墙体的建造和老旧砖墙的修复工程中。

布里斯托老维克剧院的主管汤姆·莫里斯说道："现今的布里斯托老维克剧院是一座我们为整座城市打造的剧院，而这也是250多年前该剧院建造的初衷。霍沃斯·汤普金斯建筑事务所为实现我们的雄心壮志做出了巨大的贡献。"

Haworth Tompkins have completed a new foyer and studio theater for the Grade I listed Bristol Old Vic.
After five years of research, design and construction, the new design opens up the front of house areas to more diverse audiences, placing the theater at the heart of Bristol's public life.

Bristol Old Vic – built in 1766 – is the oldest continuously working theater in the English-speaking world. The auditorium remained largely intact but successive alterations culminated in 1972 with a new foyer and studio theater by the respected architect Peter Moro. The auditorium and back of house were renovated by architect Andrzej Blonski in 2012. Recently, however, audiences became disengaged; the 1970s alterations, although skillfully realized, divided the auditorium into two unconnected foyers and sealed the theater away from the street. Radical change was required, so it was decided to replace entirely the 1972 alterations

项目名称：Bristol Old Vic / 地点：Bristol Old Vic, King St, Bristol, BS1 4ED, UK / 事务所：Haworth Tompkins / 项目团队：Beatie Blakemore, Tom Gibson, Toby Johnson, Will Mesher, Michael Putman, Steve Tompkins, Roger Watts / 总承包商：Gilbert-Ash / 剧院顾问：Charcoalblue
结构工程师：Momentum / 设备工程师：Max Fordham / 声学工程师：Charcoalblue / 工料测量师：Gardiner and Theobold
合同管理：GVA Acuity / 客户代表：Plann / 客户：Bristol Old Vic Trust / 总楼面面积：2,135m² / 造价：£9,300,000
设计开始时间：2016.10 / 竣工时间：2018.9 / 摄影师：©Philip Vile (courtesy of the architect) - p.70 right, p.71, p.73, p.75, p.78, p.79, p.82; ©Fred Howarth (courtesy of the architect) - p.74, p.76~77, p.83; courtesy of the architect - p.70 left-upper

with a more welcoming and legible space, reinstating a grand public room in the Coopers' Hall. The new foyer is an informal extension of the street, framed by structural timber and glass, allowing daylight in.

The centerpiece is the facade of the Georgian auditorium, now visible from the street, illuminated by a light-well and punctured by new openings to overwrite the evidence of historic alterations.

Architect Steve Tompkins said: "Transforming a Grade I listed building is an enormous responsibility, but I hope we have managed both to illuminate the unique history of the theater and to make a vivid new public space where anyone and everyone will feel at home."

Indeed, mezzanine galleries, winding staircases and viewing platforms offer theatergoers a single, convivial space, while locals may meet during the daytime in the foyer café-bar. A new studio theater occupies the former barrel store of the Coopers' Hall.

The south facade facing street is a public artwork consisting of moveable, hand-operated shutters which incorporate text from Garrick's inaugural address (the influential 18th Century actor and playwright) and poetry by Bristol City Poet Miles Chambers. The text highlights both the importance of the theater's history and its role in Bristol's future.

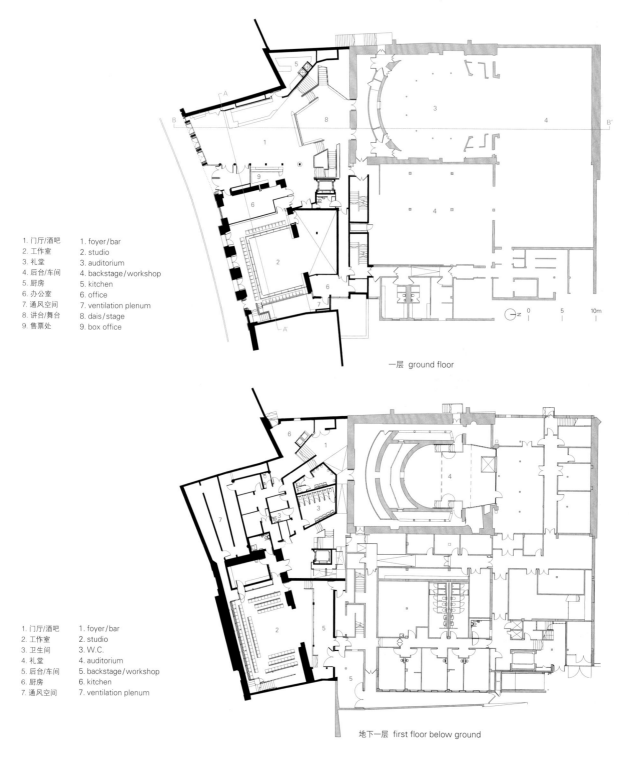

1. 门厅/酒吧　1. foyer/bar
2. 工作室　　2. studio
3. 礼堂　　　3. auditorium
4. 后台/车间　4. backstage/workshop
5. 厨房　　　5. kitchen
6. 办公室　　6. office
7. 通风空间　7. ventilation plenum
8. 讲台/舞台　8. dais/stage
9. 售票处　　9. box office

一层 ground floor

1. 门厅/酒吧　1. foyer/bar
2. 工作室　　2. studio
3. 卫生间　　3. W.C.
4. 礼堂　　　4. auditorium
5. 后台/车间　5. backstage/workshop
6. 厨房　　　6. kitchen
7. 通风空间　7. ventilation plenum

地下一层 first floor below ground

As with much of the practice's work, durable materials are chosen with capacity to mature. The Douglas Fir structure will deepen in color, an oak lath screen will bend and flex organically and a copper bar and concrete floor will become patinated with use. Detailing is direct and informal; junctions retain an outdoor language, heightening the sense that the foyer mediates between street and historic building.

The scheme uses structural timber and timber linings for new elements including frame, roof, and facade. The foyer and new studio are naturally ventilated via a large intake plenum and pre-cooling labyrinth, with thermostatically controlled high- and low-level opening windows. The moveable shutters are designed to optimize summer shading and winter solar heating. An exposed concrete floor and existing masonry surfaces throughout the space contribute thermal mass, and brickwork from demolition operations is recycled and incorporated into new walls and masonry repairs.

In the words of Artistic Director Tom Morris, "The Bristol Old Vic is now a theater for the whole city – a founding principle when it was first imagined more than 250 years ago. Haworth Tompkins has done a wonderful job in helping us to realize our ambitions."

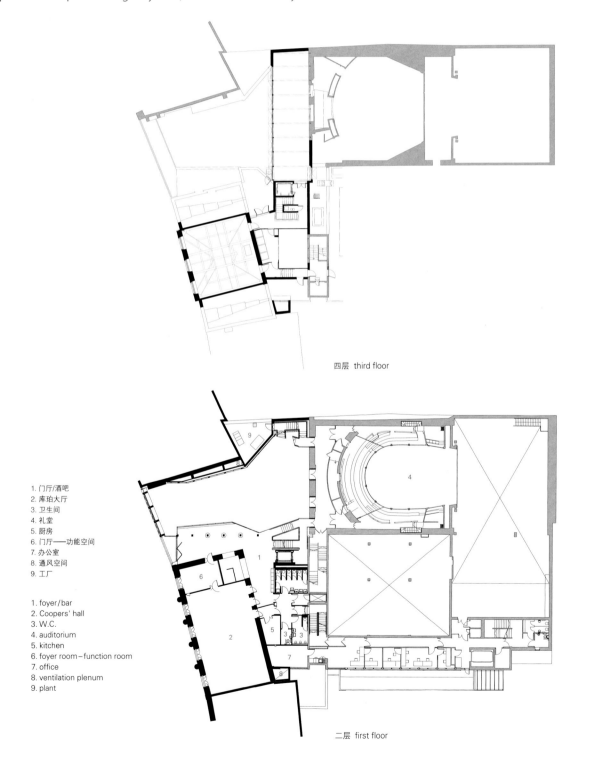

四层 third floor

1. 门厅/酒吧
2. 库珀大厅
3. 卫生间
4. 礼堂
5. 厨房
6. 门厅——功能空间
7. 办公室
8. 通风空间
9. 工厂

1. foyer/bar
2. Coopers' hall
3. W.C.
4. auditorium
5. kitchen
6. foyer room – function room
7. office
8. ventilation plenum
9. plant

二层 first floor

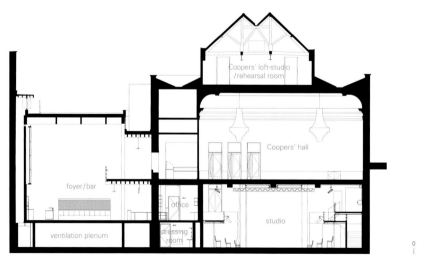

A-A' 剖面图 section A-A'

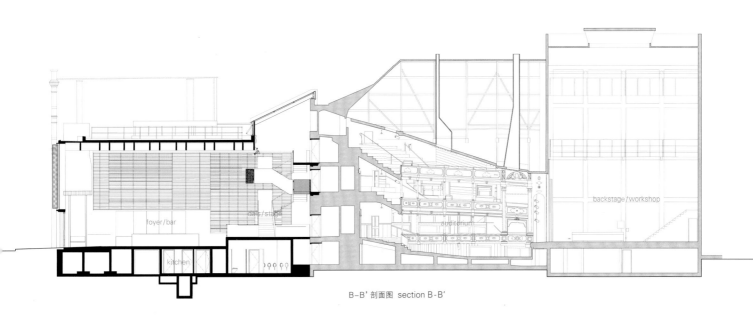

B-B' 剖面图 section B-B'

以"乌托邦"为名的图书馆和表演艺术学院已经对比利时阿尔斯特市开放了。2015年举办了一场该项目的公开竞标,之后,市政厅选择建立一种基于设计与建造合同的公私合建模式,并最终委托KAAN建筑事务所设计,Van Roey建筑公司(总承包商)建造。8000m²的砖结构建筑包括一栋引人注目的19世纪后期的历史建筑,在优雅地回应功能需求的同时又重振了城市景观。"乌托邦"现今已经成为阿卡斯特市中心的地标建筑,成为市民日常生活中争相前往的地方。

建筑名字的灵感来自于托马斯·莫尔的著作《乌托邦》。新建筑插入城市肌理中,强调了城市中心独具特色的不规则街道和私密空间,并与之互动。沿着Esplanadestraat街、Graanmarkt街和Peperstraat街有三座新广场建成。所谓的Pupillenschool,是一座1880年建造的建筑,以前是一所针对士兵子女的学校。当时,孩子们在16岁的时候可以登记服兵役,在16岁前他们就在这所学校接受教育。这座建筑也被纳入KAAN建筑事务所的设计中,作为新建筑的基石。

"乌托邦"与城市及居民密不可分,大楼的入口位于阅读咖啡厅和礼堂之间的小广场上。在宽敞的大厅中穿行,就会看到建筑内的开放式景观从地面一直延伸到天花板,数块厚实的混凝土楼板在空间中悬挑出来,如同漂浮着一般。每一块都悬挂在不同的高度上,每层楼均设有书架和阅读桌,楼板指向中庭和原有建筑的砖墙立面。此外,一个11.5m的书柜伸向天花板,并装满了阿尔斯特居民捐赠的书籍。混凝土结构如同被书籍支撑着,书柜支撑起混凝土板,使得楼板悬挑伸出而无需额外的支撑。楼梯呈锯齿状向上延伸,在华丽的中庭和阅览室周围如雕塑般存在。与此同时,天花板已经被简化到几乎无法察觉的地步。拉伸的金属色网格遮挡住了所有的技术系统,同时柔和了强烈的日光并在白天营造出愉快的氛围。

礼堂设在一层,表演艺术学院设在一层和二层,位于阅读中庭两侧。新建筑内的芭蕾舞室、排练室和教学场所的窗户与房间本身一样高大宽阔,这些窗户是从砖结构中精致切割出来的,不但能让室内的

乌托邦——图书馆和表演艺术学院
Utopia – Library and Academy for Performing Arts
KAAN Architecten

阿尔斯特城市肌理与地标
Aalst urban fabric and landmarks

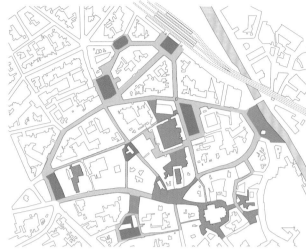

阿尔斯特街道与广场城市网络
Aalst urban network of streets and squares

人们饱览城市景观,还能让城市中的人看到室内的情况,同时也是立面的组成部分。旧建筑也使用了相同的表达语言,Pupillenschool以前的窗户栏杆被移除,钢琴室的窗台也被降低了不少。

建筑的外墙大部分由新砖砌成。在对阿尔斯特市的主要颜色进行研究后,建筑师选择了名为"阿尔斯特红"的深色砖。为了突出"乌托邦"的双重特性,这些长而薄的砖(50cm×10cm×4cm)以水平方式铺设,以此来平衡垂直铺设的旧学校立面。无论是在内部还是在外部,历史立面都与宽敞的空间实现了完美的融合,砖体本身也与淡灰色的混凝土构件形成了对话效果。

对于建筑师来说,声学是一个基本的设计元素:在图书馆阅读不能受到音乐课和排练的干扰。门板和替代旧建筑中木地板的悬挑混凝土楼板被改造成了隔声屏障,而双层玻璃窗能捕获每一个钢琴音符。

"乌托邦"在注重开放性的同时也强调可持续性。该建筑获得了BREEAM优秀等级认证:建筑材料和劳动力都在当地解决,施工过程中使用低能耗机器,太阳能电池板、地热、LED照明都被集成在设计中,对雨水进行再利用与收集,被拆下来230 000块砖块在其他地方得到了重新使用。

Utopia, the Library and Academy for Performing Arts has opened its doors to the Belgian city of Aalst. Following an open competition in 2015, the City Council chose to establish a Public-Private Partnership (PPP) based on a Design & Build contract and assigned the project to KAAN Architec-

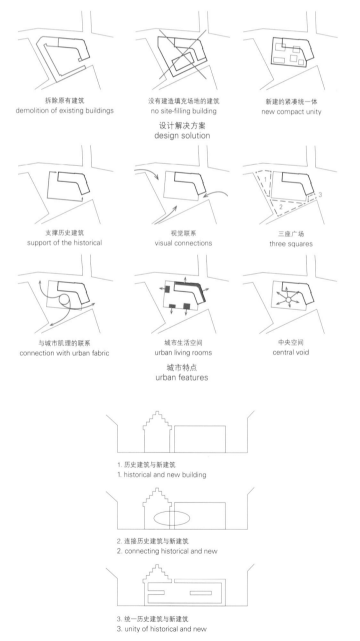

ten (architect) and Van Roey (main contractor). The 8,000m² brick structure incorporates a strikingly historic building from the second half of the 19th century and rejuvenates the urban landscape while elegantly accomplishing the required functionality. Utopia has already become a new landmark in Aalst city center that citizens are eager to enjoy in their everyday lives.

Taking a cue from Thomas More's acclaimed book *Utopia*, the new building has been slotted into the urban fabric to enhance and interact with the characteristic irregular streets and intimate spaces of the city center. Three new squares have been created alongside Esplanadestraat, Graanmarkt and Peperstraat. The so-called Pupillenschool from 1880, a former school where children of soldiers were educated until the age of 16 when they could register for a regiment, has been embedded into KAAN Architecten's design as the cornerstone of the new building.

Utopia, the city, and its residents are inextricably linked. The entrance to the building is located on an intimate square between the reading café and the auditorium. Moving through the wide hall, the open interior landscape unfolds from floor to ceiling; several thick concrete floors are cantile-

南立面 south elevation

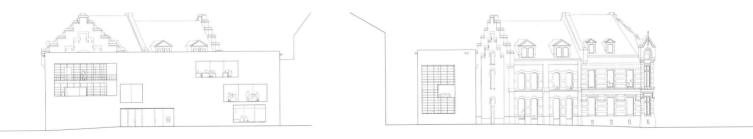

东立面 east elevation　　　　　　　　西立面 west elevation

北立面 north elevation

1. 礼堂
2. 餐厅
3. 咨询台
4. 主阅览厅
5. 儿童阅读剧院
6. 儿童阅读区
7. 音乐与视频收藏区
8. 艺术文献收藏区
9. 音乐教室
10. 剧院教室
11. 青年工作室

1. auditorium
2. restaurant
3. information desk
4. main reading hall
5. children reading theater
6. children reading area
7. music and video collection
8. art literature collection
9. music classroom
10. theater classroom
11. youth atelier

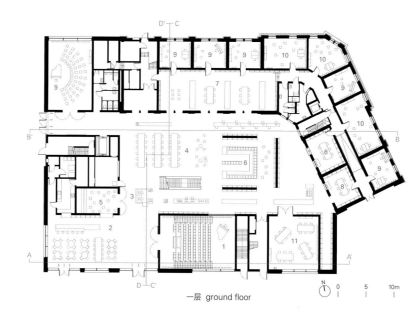

一层 ground floor

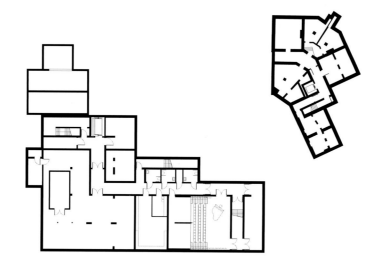

地下一层 first floor below ground

1. 餐厅
2. 礼堂
3. 青年工作室
4. 芭蕾舞室
5. 理论教室
6. 计算机区

1. restaurant
2. auditorium
3. youth atelier
4. ballet room
5. theory classroom
6. computer area

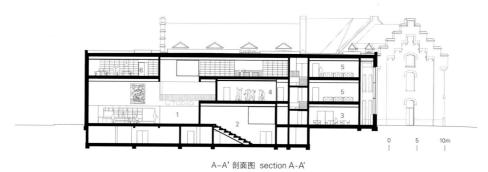

A-A' 剖面图 section A-A'

1. 剧院教室
2. 艺术文献收藏区
3. 主阅览厅
4. 办公室
5. 音乐教室

1. theater classroom
2. art literature collection
3. main reading hall
4. offices
5. music classroom

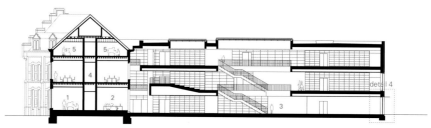

B-B' 剖面图 section B-B'

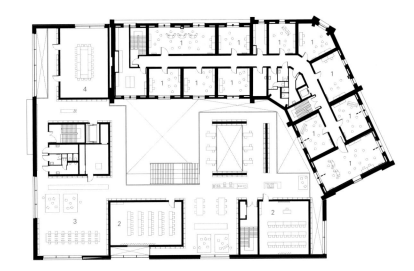

1. 音乐教室
2. 理论教室
3. 计算机区
4. 会议室

1. music classroom
2. theory classroom
3. computer area
4. meeting room

三层 second floor

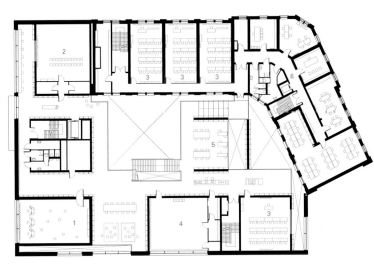

1. 音乐教室
2. 剧院教室
3. 理论教室
4. 芭蕾舞室
5. 主阅览厅
6. 办公室

1. music classroom
2. theater classroom
3. theory classroom
4. ballet room
5. main reading hall
6. offices

二层 first floor

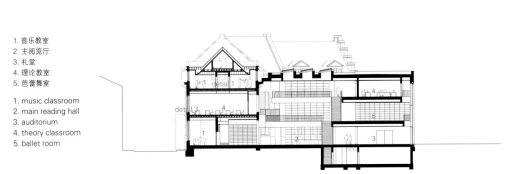

1. 音乐教室
2. 主阅览厅
3. 礼堂
4. 理论教室
5. 芭蕾舞室

1. music classroom
2. main reading hall
3. auditorium
4. theory classroom
5. ballet room

C-C' 剖面图 section C-C'

1. 餐厅
2. 咨询台
3. 音乐与视频收藏区
4. 音乐教室
5. 理论教室

1. restaurant
2. information desk
3. music and video collection
4. music classroom
5. theory classroom

D-D' 剖面图 section D-D'

vered into the space as if they are floating. Hanging at varying heights, each level features bookshelves and reading tables, while looking into the atrium and towards the brick facade of the pre-existing building. Moreover, an 11.5m-high bookcase, filled with books donated by Aalst residents, stretches towards the ceiling that the concrete structures seem to be supported by books. The bookcases are pushed up against concrete discs which allow the floors to cantilever out without extra support. Mimicking the treads, the stairs zig-zag upwards, giving the staircase a sculptural presence at the periphery of the magnificent atrium and reading room. The ceilings have been minimalized to the point of being almost undetectable. All the technical systems are concealed behind a stretched metal-colored mesh that softens the strong daylight and creates a pleasant atmosphere during the day.

Apart from the auditorium on the ground floor, the Academy for Performing Arts is housed on the first two floors, flanking the reading atrium. Within the new building, the ballet room, rehearsal studios, and teaching spaces have windows as tall and wide as the rooms themselves, carefully sliced out of the brickwork, providing a view to the city and allowing a glimpse from the city, while contributing to the facade composition. Using the same expressive language,

the railings of the former Pupillenschool's windows have been removed and the windowsills of the piano nobile have been significantly lowered.

Much of the building's exterior are new brickwork. After studying the predominant colors of Aalst, the architects chose a dark brick called "Red Aalst". To accentuate Utopia's duality, these long flat bricks (50cm x 10cm x 4cm) have been laid horizontally to complement the vertically-oriented old school facades. Both outside and inside, the historic facades blend perfectly with the generous spaces, while the brickwork dialogues with light gray concrete elements.

Acoustics is a fundamental design tool for the architects: the reading in the Library should not be disrupted by music lessons and performance rehearsals. The doors and the suspended concrete floors that replace the original wooden floors are transformed into sound barriers while double glazed windows capture each single piano note.

Utopia's openness also exudes sustainability. The building has achieved a BREEAM Excellent rating: materials and labour are locally sourced; low-energy machines are used for construction; solar panels, geothermal heat and LED lighting are integrated in the design; rainwater is recuperated and buffered; 230,000 bricks are chipped away and reused elsewhere.

项目名称：Utopia–Library and Academy for Performing Arts
地点：Utopia 1, Aalst, Belgium
事务所：KAAN Architecten
主创建筑师：Kees Kaan, Vincent Panhuysen, Dikkie Scipio
项目团队：Bas Barendse, Tjerk de Boer, Sebastiaan Buitenhuis, Sebastian van Damme, Paolo Faleschini, Raluca Firicel, Narine Gyulkhasyan, Joost Harteveld, Walter Hoogerwerf, Martina Margini, Giuseppe Mazzaglia, Kevin Park, Giulia Rapizza
总承包商：Groep Van Roey NV, Rijkevorsel
建议咨询：UTIL Struktuurstudies, Schaarbeek
建议技术设施、水电设施：Studiebureau R. Boydens NV, Brugge
防火：ABT, Delft
声学设计：Tractebel Engineering SA, Brussels
可持续性设计：Studiebureau R. Boydens NV, Brugge Sint-Michiels
客户：Autonoom Gemeentebedrijf Stadsontwikkeling Aalst (AGSA)
总楼面面积：8,309m² + 235m² (bike parking)
设计开始时间：2016.3
施工时间：2016.8—2018.5
摄影师：©Delfino Sisto Legnani e Marco Cappelletti (courtesy of the architect)

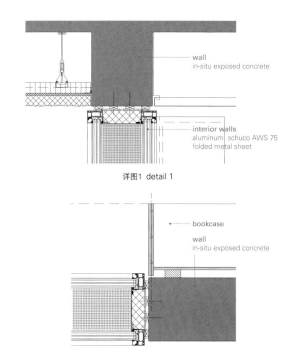

详图1 detail 1

详图2 detail 2

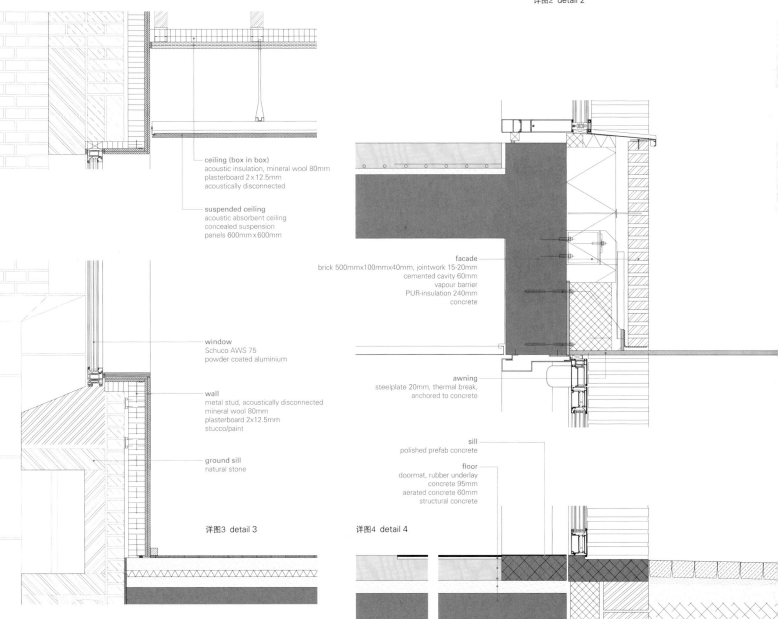

详图3 detail 3

详图4 detail 4

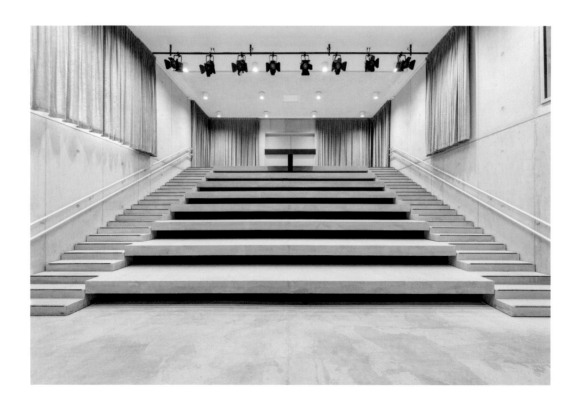

Musis Sacrum 音乐厅
Musis Sacrum
van Dongen-Koschuch

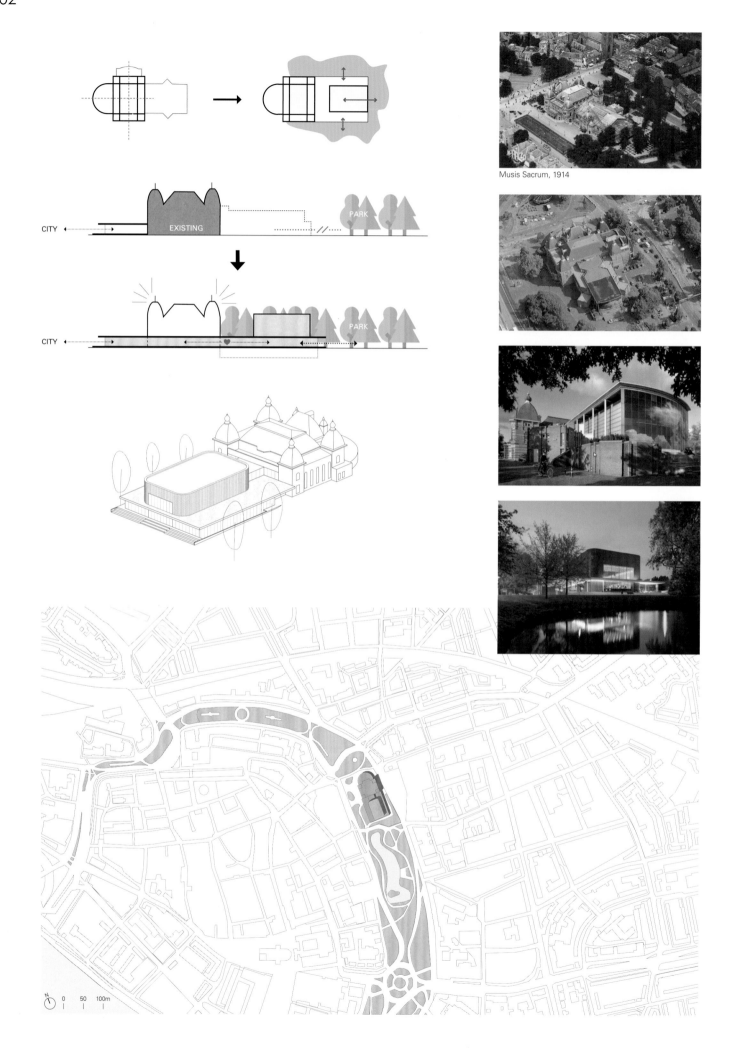

Musis Sacrum, 1914

位于荷兰阿纳姆的Musis Sacrum项目涉及对一座现存的城市公园的重新设计，对一座原有的纪念性建筑进行改造处理以及出于长久使用的目的，对一个已建立的机构进行明确的加建处理。那些来自阿纳姆及周边地区欣赏音乐和剧场表演的主体人群就是Musis & Stadstheater这一建筑混合体的使用者。因此，该项目的主要目的在于为客户建立文化项目的配套设施，以及为荷兰先锋古典剧团之———"Het Gelders Orkest"建造一个温馨的家。

自1847年来，Musis Sacrum音乐厅及其附属建筑就呈现出不同的形式和姿态，应对着不同时代的短期需求。van Dongen-Koschuch建筑事务所的解决方案的特色在于额外建立一个大型音乐厅和相关配套设施。多功能化的设计和经过深思熟虑的声学系统让Musis Sacrum音乐厅成为所有种类演出和活动的场地，包括从交响乐到流行乐等所有类型的音乐会。

与之前进行的扩建不同，新的多用途音乐厅被设计为独立于原有建筑的单独体量而存在于公园之中，建筑师将扩建部分设计为一个充满吸引力的透明体量，在优美的环境中与原有历史建筑呼应。新建的大型音乐厅的舞台背后有一面朝向公园景色的巨大玻璃窗，既可以为表演提供美丽的自然背景，又可以完全打开作为室外表演空间。正是这种独到的设计，使得Musis Sacrum音乐厅可以承办开放式的公众活动，在阿纳姆市中心开启公园的最优化使用模式。

建筑师的设计灵感来源于Musis Sacrum音乐厅最初的设计原则：在自然美景和极佳的声学条件下表演和欣赏音乐。该设计清晰地呈现出公园、Musis Sacrum音乐厅和Gelders Philharmonic交响乐团及原有音乐厅的特征。

除了对公园进行重新设计，该建筑项目还包括两项主要的介入性设计。

一、新扩建：所有的战后加建建筑都将被拆除，从而创造出一个庞大、高效以及多功能的实体空间，来承办各类与音乐相关的活动。这一新建筑的主体事实上是设置在地下的，这样就将建筑对公园造成的视觉冲击降到最低。结合超清晰玻璃幕墙、生动的屋顶以及充满自然韵味的陶瓷立面的大量使用，新加建建筑与周围的环境和谐地融合了在一起。

二、对旧的历史性建筑进行复原性的改造，使其符合当今流行的热情、合理以及安全的标准。这样设计的目的在于通过将所有不美观的技术方面的必要设施隐藏起来，恢复历史性建筑往日的辉煌。因此，项目预算方面的支出并不是立即可见的：古老的天花板、洒水器和散热器都被拆除了，墙面的裂纹得到了修复，原有的木质地板进行了重新恢复，东面的立面在原来一幅1888年的绘画作品的基础上进行了重建。最初的拱和天花板的装饰如今遍布整个音乐厅，让这座世界闻名、人们顶礼膜拜的音乐场馆重新焕发往日的光芒。

The Musis Sacrum project – in Arnhem, the Netherlands – involved redesigning an existing city park, renovating an original monument and creating a definitive new addition to an established institution with long-term use in mind. The users of the building complex, Musis & Stadstheater, are the ones that enjoy the main facilitator of music and theater performances in Arnhem and its surroundings. So the aim of the project is to facilitate the client's cultural program, as well as to create a home for "Het Gelders Orkest", one of the Netherlands' leading classical orchestras.

Since 1847 the concert halls of Musis Sacrum – including various extensions – had taken on a range of different forms and positions, each of which answered only the short-term needs at the time. Architects van Dongen-Koschuch's solution features an additional large concert hall and supportive functions; the multi-functional design and carefully designed acoustics enable the Musis Sacrum to continue as home to its wide range of shows and events, which range from symphony concerts to pop gigs and everything in between.

项目名称：Musis Sacrum / 地点：Arnhem, The Netherlands / 建筑师：van Dongen-Koschuch / 首席建筑师：Frits van Dongen, Patrick Koschuch / 设计团队：Ralph van Mameren, Elisabetta Bono, Rui Duarte, Hesh Fekry, Maikel Super, Casper de Heer, Daan Vulkers, Klaas Sluijs, Olga Moreno / 承包商：Mertens Bouwbedrijf bv, Weert & Homij, Groningen / 剧院设计建议：Theateradvies bv, Amsterdam / 建筑物理&声学设计：Peutz bv, Zoetermeer / 设备设计建议：Nelissen Ingenieursbureau bv, Eindhoven / 结构工程师：Aronsohn/VandeLaar / 客户：Gemeente Arnhem, Musis & Stadstheater Arnhem / 玻璃立面：Metaglas BV / 陶瓷立面：Koninklijke Tichelaar / 室内组件：Vola, FSB, Not Only White, Kone elevators, Merford firedoors, Mosa tiles, Espero / 家具：Lensvelt, Vitra, VeryWood, Arper, Jezet Bronze components: Aurubis / 用地面积：3,700m² / 总楼面面积：9,820m² (new 5,950m² + existing building renovation 3,870m²) / 造价：13.6M / 竣工时间：2018
摄影师：©Bart van Hoek (courtesy of the architect)

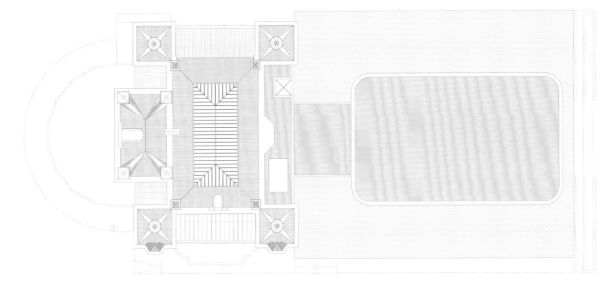

屋顶 roof

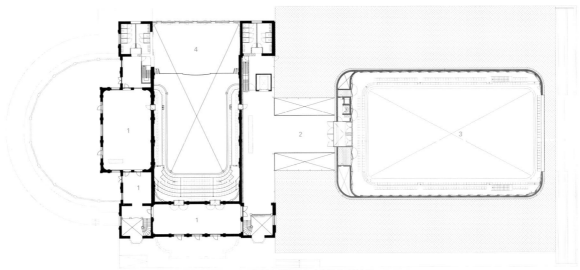

1. 门厅 2. 新门厅 3. 主大厅 4. 古典大厅
1. foyer 2. new foyer 3. main hall 4. classic hall
二层 first floor

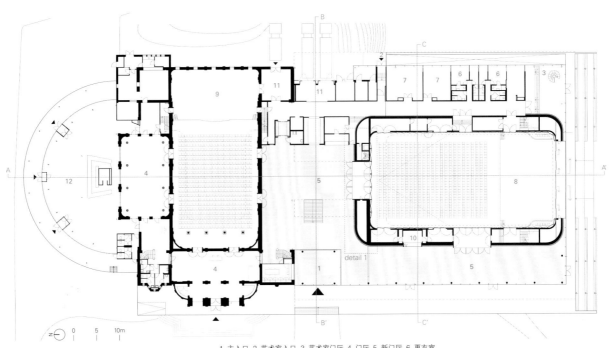

1. 主入口 2. 艺术家入口 3. 艺术家门厅 4. 门厅 5. 新门厅 6. 更衣室
7. 办公室 8. 主大厅 9. 古典大厅 10. 酒吧 11. 装载区 12. 餐厅
1. main entrance 2. artist entrance 3. artist foyer 4. foyer 5. new foyer 6. changing room
7. office 8. main hall 9. classic hall 10. bar 11. loading area 12. restaurant
一层 ground floor

Unlike any former extension, the new multi-purpose hall is positioned as a separate volume in the park; it is an inviting and transparent pavilion which respectfully complements the historic building and beautiful surroundings. This new concert hall features a large glass window behind the main stage that not only acts as a botanical backdrop, but also can be opened up onto the park for outdoor performances. Thanks to this unique feature, Musis Sacrum can offer open-air public events, making optimal use of a wonderful park in the heart of Arnhem.

The architects draw inspiration from the original principle of the Musis Sacrum Institute: to perform and listen to music in excellent acoustic conditions within an attractive green context. The design attempts to reflect the characters of the park, the Musis Sacrum Institute, the Gelders Philharmonic Orchestra and the identity of the original concert hall. Besides redesigning the park, the architectural project comprises two main interventions.

Firstly, the new extension: all post-war additions are demolished to create the large, efficient and multi-functional entity which facilitates all music-related events. The majority of this new building is in fact underground to minimize the visual impact on the park. Together with the extensive use of ultra-clear glass, a living roof, and a ceramic facade in natural tones, the new addition blends comfortably with its surroundings.

The second part of the project consists of restoring the old monument, bringing it up-to-date with current fire, sound and safety standards. The aim was to restore the monument to its former glory by hiding all unsightly technical necessities from view. As a result, much of the budgetary spent on this operation is not immediately visible: old ceilings, sprinklers and radiators are removed, cracks in the walls repaired, original wooden floors restored and the east facade rebuilt from an original 1888 drawing. The original arches and ceiling ornamentation are now consistent throughout the concert hall, bringing back the former grandeur of a well-known and highly-respected musical arena.

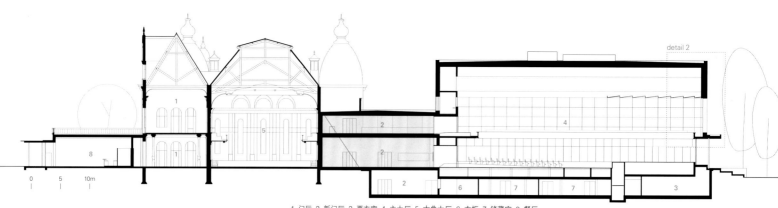

1. 门厅 2. 新门厅 3. 更衣室 4. 主大厅 5. 古典大厅 6. 衣柜 7. 储藏室 8. 餐厅
1. foyer 2. new foyer 3. changing room 4. main hall 5. classic hall 6. wardrobe 7. storage 8. restaurant
A-A' 剖面图 section A-A'

1. 主入口 2. 新门厅 3. 厨房 4. 装载区
1. main entrance 2. new foyer 3. kitchen 4. loading area
B-B' 剖面图 section B-B'

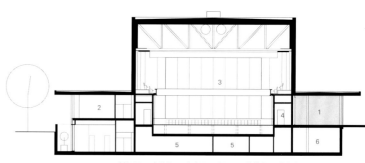

1. 新门厅 2. 办公室 3. 主大厅 4. 酒吧 5. 储藏室 6. 设备间
1. new foyer 2. office 3. main hall 4. bar 5. storage 6. installations room
C-C' 剖面图 section C-C'

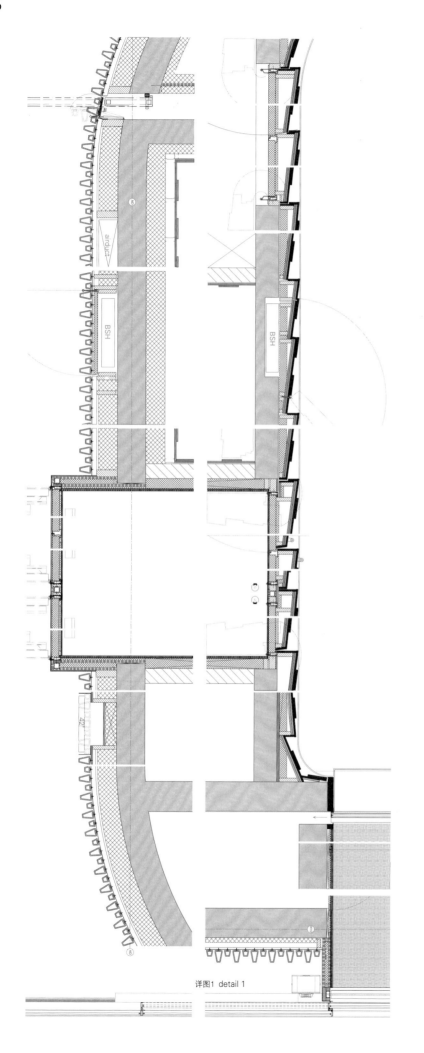

详图1 detail 1

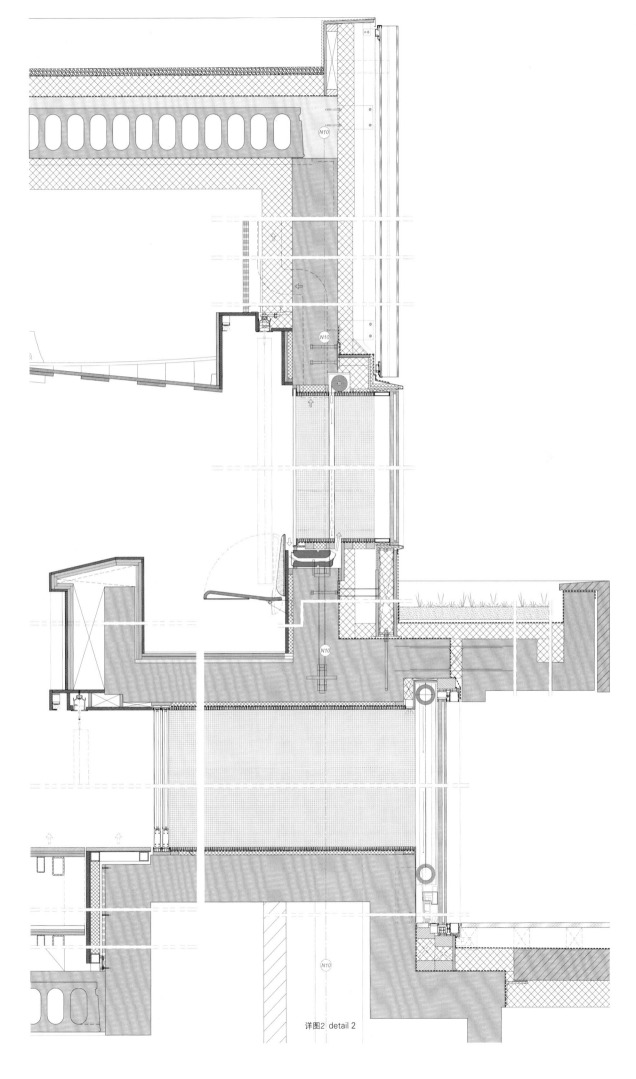

详图2 detail 2

Coal Drops Yard 购物中心
Coal Drops Yard
Heatherwick Studio

长期以来，Heatherwick工作室一直在国王十字车站区域办公，现如今它已将两座19世纪50年代的传统铁路建筑改造成一个新的商业区。

2014年，国王十字中央有限合作公司委托Heatherwick工作室对该地区进行了彻底的重新规划。这两座形状狭长的维多利亚式建筑最初的用途是接收采自英格兰北部的煤炭，然后通过驳船和卡车向伦敦周边地区进行分配。但随着时间的推移，华丽的铸铁和砖结构变得越来越不美观，渐渐被遗弃，在20世纪90年代被彻底遗弃之前它变成了为轻工业、仓储和夜总会服务的设施。

工作室面临的挑战在于将破旧的建筑和长而棱角分明的场地改造成一个活跃的零售区，公众可以在这里聚集和走动。

Heatherwick工作室的设计扩展了内部的山墙式仓库屋顶，将两条高架桥连接到一起，从而形成了一个院子的概念，并创造了流动的交通模式。流动式的屋顶由一个蕴含于传统构造之内的、全新的、充满高技术含量的、独立的结构支撑着。屋顶在边缘继续上升，伸展并最终实现了接触。这就形成了一个全新的悬浮式的顶层设计，一个巨大的有顶室外空间和一个场地中心聚集区。

从高处的有利位置，游客可以欣赏周围的景色。在底部，人们可以在一个有遮蔽的20m高的空间里逗留，这里也适合作为举办音乐会和表演的场所。

屋顶的形状和颜色融合了新旧风格，非常符合该地区的风格。35m宽的结构复杂的扩建部分从原来的山墙开始延伸，创造出两栋建筑在空中轻轻接触的错觉。自支撑结构包括52根钢柱，它们隐藏在老化的砖和铁结构里面，由混凝土墙和核心筒支撑，以保持传统构件的完整性。

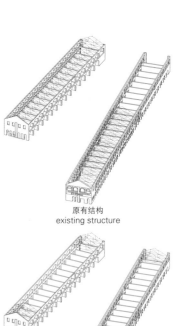

原有结构
existing structure

可调节木屋顶桁架
adjusted timber roof truss

新建自支撑钢结构与两个新建钢屋顶剥离式桁架
new self-supporting steel structure
and two new steel roof peel trusses

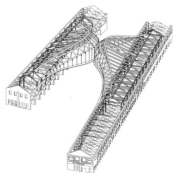

新建上层楼板，悬挂于屋顶剥离式桁架
new upper level floor hung from roof peel trusses

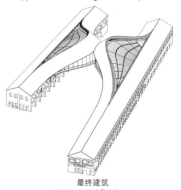

最终建筑
architectural finishes

每个弯曲的带状结构的新屋顶都由20个钢构件组成,这些钢构件用螺栓固定在四根与柱子相连的桁架上。顶层全景的框架是由64块交错的全高结构玻璃板组成的。屋顶覆盖层包括80 000多块来自同一个威尔士采石场的蓝灰色石板瓦,从而与原来的维多利亚式建筑保持了纹理上的一致。

在试图改善和改造原有历史建筑时,工作室采用了微调的手法。所有必要的加建结构都用上了能够体现出旧铁、烟灰砖、石板、木板和鹅卵石这些材质的颜料。与此同时,设计还保留了最近的标记和墙面上乱涂乱画的痕迹,这些丰富的纹理设计都保留了最初的Coal Drop时期煤炭的独有特征。

零售区总面积为9300m²,用于购物、餐饮和活动的举办,这里由一系列相互连接的街道组成,内含55个大小各不相同的单元,容纳众多零售商品,包括小摊位和知名品牌大商铺。

入口分布于Stable大街和高架桥的两端,通过桥梁和楼梯等多个连接处也能够进入,无障碍空间使人们可以自然地行走。这一具有渗透性和独特性的公共空间有助于将国王十字大街更广泛地转化为充满活力的伦敦街区。

Heatherwick工作室的创始人托马斯·希瑟威克说:"这些令人惊叹的维多利亚式建筑最初不是为数百人居住而建造的,而是伦敦封闭式基础设施的一部分。这么多年来,它一直发挥着不同的服务功能,现如今,我们很高兴有机会利用我们的设计理念来完全开放这一区域,创造出新的空间,让每个人都能体验到这些丰富而有特色的建筑。"

As a long-time resident of King's Cross, Heatherwick Studio has reinvented two 1850s heritage rail buildings as a new shopping district.

In 2014, King's Cross Central Limited Partnership commissioned the studio to radically rethink the site. The pair of elongated Victorian buildings originally received coal from Northern England for distribution around London by barge and cart. But with time the ornate cast-iron and brick structures became increasingly derelict, serving as light industry, warehousing and nightclubs before abandonment in the 1990s.

The challenge was to transform the dilapidated buildings and long, angular, site into a lively retail district where the public could gather and circulate.

Heatherwick's design extends the inner gabled warehouse roofs, linking two viaducts to define a yard, creating fluid circulation patterns. The flowing roofs – supported by a new, highly technical, freestanding structure interlaced within the heritage fabric – rise, stretch and eventually touch. This forms a new floating upper storey, a large covered outdoor

1. 零售店 2. 次级垂直交通流线 3. 主垂直交通流线
1. retail 2. secondary vertical circulation 3. primary vertical circulation
一层 ground floor

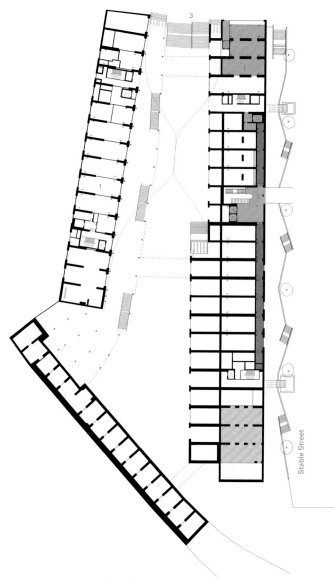

1. 零售店 2. 次级垂直交通流线 3. Cubitt广场 4. 服务走廊
1. retail 2. primary vertical circulation 3. Cubitt square 4. service corridor
夹层 mezzanine

南立面 south elevation

东立面——东Coal Drops east elevation _ East Coal Drops

东立面——西Coal Drops east elevation _ West Coal Drops

1. 零售店 2. 次级垂直交通流线 3. 主垂直交通流线 4. Cubitt广场
1. retail 2. secondary vertical circulation 3. primary vertical circulation 4. Cubitt square
二层 first floor

1. 零售店
1. retail
三层 second floor

北立面 north elevation

西立面——东Coal Drops west elevation _ East Coal Drops

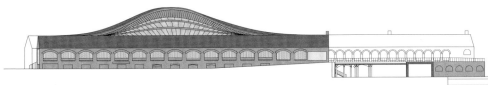

西立面——西Coal Drops west elevation _ West Coal Drops

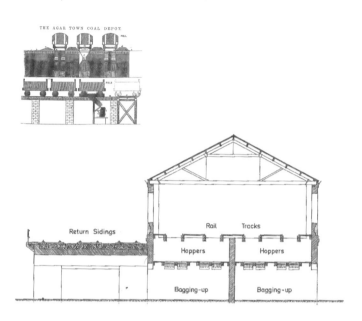

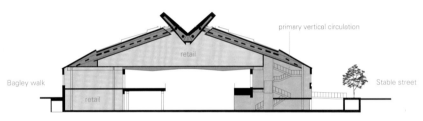

A-A' 剖面图 section A-A'

B-B' 剖面图 section B-B'

space and a central focus for the site.

From the elevated vantage point, visitors enjoy views of the surrounding locale. Below, people may linger in a sheltered twenty-meter-high space, also suitable as a concert and performance venue.

Amalgamating old and new, the roof form and patina are site-specific. The 35m-wide structurally complex extension flows from the original gables, creating the illusion of two buildings lightly touching in mid-air. The self-supporting intervention comprises 52 steel columns concealed behind aged brick and iron, shored up by concrete walls and cores, to preserve the integrity of heritage elements.

Each new curving roof ribbon is formed by 20 steel sections bolted onto four trusses tied to the columns. Framing the top floor's panoramic outlook are 64 staggered panels of full-height structural glass. The roof cladding includes over 80,000 blue-gray slate tiles from the same Welsh quarry as the original Victorian building, giving a consistent hue. When seeking to enhance and adapt the existing historic buildings, the studio adopted a light touch. All necessary additions drew on the palette of aged ironwork, soot-stained brick, slate, timber boards and cobbled stone setts. Also retaining more recent signage and graffiti, these rich textures preserve the Coal Drops' distinct character.

Totalling 9,300m² for shopping, dining and events, the retail quarter is a series of interlinked streets: the 55 units vary in size to accommodate a range of retailers – fledgling pop-ups and large-scale units for established brands.

With entrances scattered along Stable Street and at both ends of the viaducts, as well as multiple connections into the yard via bridges and stairs, the accessible space encourages people naturally to circulate. This permeable and distinctive public space contributes to the wider transformation of King's Cross into a vibrant London district.

Thomas Heatherwick, Founder of Heatherwick Studio, said: "These amazing Victorian structures were never originally built to be inhabited by hundreds of people, but instead formed part of the sealed-off infrastructure of London. After serving so many varied uses throughout the years, we've been excited by the opportunity to use our design thinking to finally open up the site, create new spaces and allow everyone to experience these rich and characterful buildings."

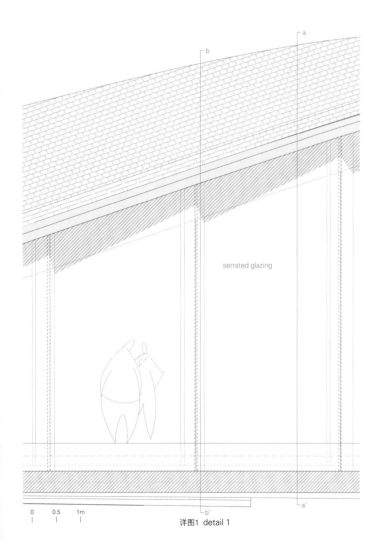

详图1 detail 1

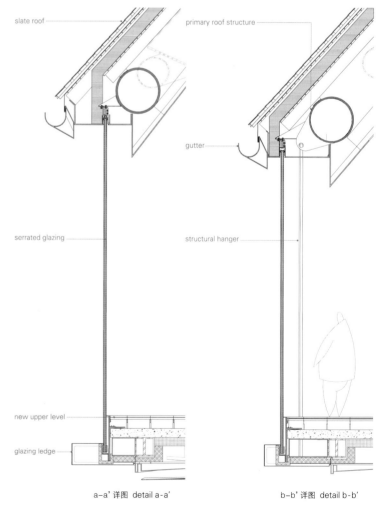

a-a' 详图 detail a-a'   b-b' 详图 detail b-b'

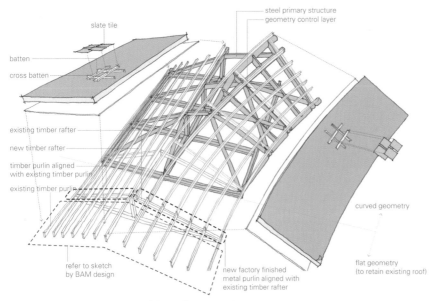

板条屋顶详图 slate roof detail

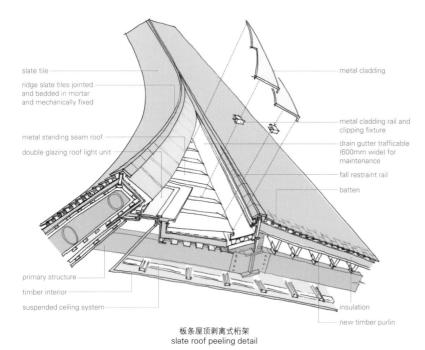

板条屋顶剥离式桁架
slate roof peeling detail

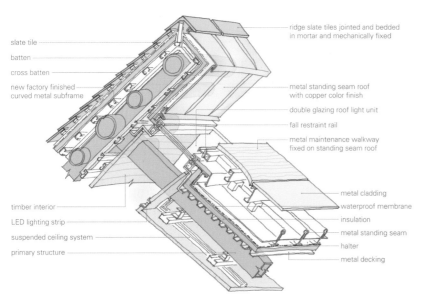

剥离式屋顶——屋顶入窗剖面详图
roof peel - roof light sectional detail

项目名称：Coal Drops Yard
地点：King's Cross, London, UK
事务所：Heatherwick Studio
设计指导：Thomas Heatherwick
团队负责人：Lisa Finlay / 项目负责人：Tamsin Green
项目团队：Jennifer Chen, Andrew Edwards, Daniel Haigh, Phil Hall-Patch, Steven Howson, Sonila Kadillari, Michael Kloihofer, Nilufer Kocabas, Elli Liverakou, Ivan Linares Quero, Mira Naran, Ian Ng, Thomas Randall-Page, Emmanouil Rentopoulos, Dani Rossello Diez, Angel Tenorio, Takashi Tsurumaki, Pablo Zamorano
制作团队：Jordan Bailiff, Einar Blixhavn, Erich Breuer, Darragh Casey, Ben Dudek, Alex Flood, Freddie Lomas, Hannah Parker, Monika Patel, Luke Plumbley, Jeff Powers
客户：Argent LLP / 开发商：KCCLP/Argent LLP
遗迹咨询：Giles Quarme & Associates
结构、立面工程师：Arup / 机电、可持续性：Hoare Lea
照明设计师：Speirs and Major
成本顾问：Gardiner and Theobald
交付建筑师：BAM Design / 板条生产：Welsh Slate Ltd
用地面积：13,600m² / 总室内面积：13,935.456m²
总零售空间面积：9,290.304m²
设计时间：2014—2018 / 施工时间：2016—2018.10
摄影师：©Hufton+Crow (courtesy of the architect) - p.122~123, p.129[upper], p.130~131; ©Luke Hayes (courtesy of the architect) - p.118~119, p.126[lower], p.127[top], p.129[lower], p.133; ©John Sturrock (courtesy of the architect) - p.126[upper], p.127[bottom-left, bottom-right]

红十字会志愿者之家
Red Cross Volunteer House
COBE

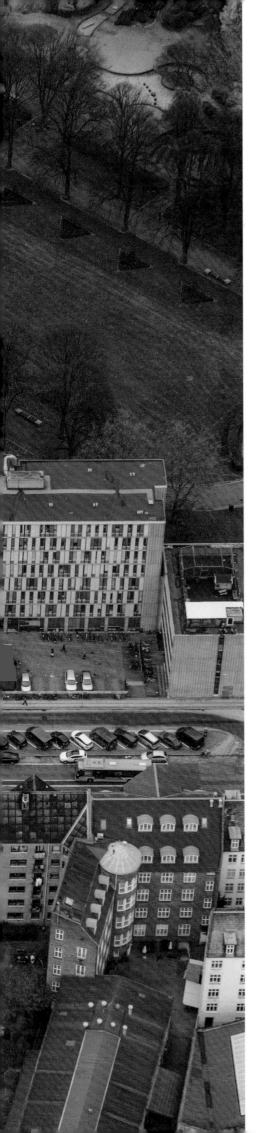

红十字会志愿者之家是丹麦哥本哈根红十字会国家总部的扩建建筑,为34 000名红十字志愿者提供一个工作持续发展的场所。红十字会志愿者之家项目是由COBE建筑事务所与志愿者们合作设计完成的,体现出对志愿者工作的肯定,也是工作人员、志愿者和红十字会中愿意为边缘化的市民做出贡献的人们的聚会场所。

客户的想法是要创造出一片新的开放空间——一处位于城市中的富有活力的区域,以此回馈这座城市。建筑师、COBE建筑事务所创始人丹·斯图贝尔嘉德解释说:"我们希望通过红十字会志愿者之家的设计,创造一个最佳的环境献给我们的平民英雄——那些为帮助边缘化的市民做出了非凡努力的成千上万的志愿者。"

三角形的建筑拥有850m²的屋顶空间,这里被设计成一个从一层延伸至建筑物三层的大型公共台阶区域。这一独特的台阶也扮演着一个阶梯平台的角色,成为吸引员工、志愿者甚至是路人使用的充满吸引力的聚会场所。今天,这里已经成为非常受欢迎的聚会地点。在夏天,志愿者和工作人员可以在这里举行会议,而当地人则可以在台阶上进行锻炼或者喝杯咖啡等等。

红十字会志愿者之家项目是由COBE建筑事务所与志愿者代表合作开发的,从空间的规划到建筑的设计,都经过了精密的测试。在这个过程中,各方通过对话讨论和修改方案,最终取得巨大的成功。2017年11月建筑开放后,红十字会吸引了更多的游客,当地其他的分支机构都会到这里参观,数不胜数的充满好奇心的游客也会在这里驻足游览。

正如丹麦红十字会秘书长安德斯·拉德卡尔所描述的那样,"这座新建筑具有开放和诱人的建筑风格,符合我们与社区建立对话的雄心壮志。"

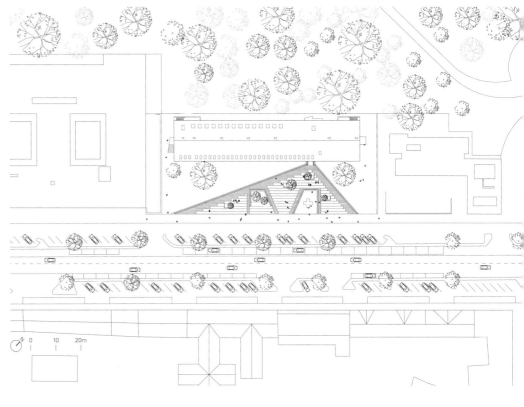

独一无二的坡屋顶设计使得建筑与该地区独特的风格融为一体，与附近的丹麦共济会大厅、Fælledparken公园以及曾经是哥本哈根市政厅大楼但现如今作为红十字会总部的黄色砖制的大楼一道重新诠释了地区特色。而志愿者之家的台阶式的屋顶所用的材料也是同样的黄色的砖。

扩建建筑的室内空间面积为750m²，部分位于地下，并作为志愿者中心和总部共享的主入口区域。左侧是志愿者之家一个可容纳100人的礼堂，这个地方通向一系列可以用于培训、会议、公开活动、演讲和放映电影的会议室。

台阶式的屋顶朝向总部开放，在志愿者之家的任何位置都能够看到这个地方。原有建筑与扩建建筑通过一座绿化公园进一步连接起来。红十字会志愿者之家也为员工和志愿者们提供了一个可以见面的共有空间，从而为蓬勃发展与合作提供了理想的条件。正如丹·斯图贝尔嘉德在谈话中所提到的那样："这栋建筑成为一个城市空间，谦逊而慷慨地表现着自我，并邀请外部世界进入其中。"

The Red Cross Volunteer House in Copenhagen, Denmark, is an extension of the organization's national headquarters, and gives 34,000 Red Cross volunteers a setting for the continuous development of their work. The building was designed by COBE in cooperation with the volunteers; it was intended both as a celebration of their commitment and as a meeting place for staff, volunteers, and anyone

wishing to contribute to the work which the Red Cross does for the city's marginalized citizens.

The client's ambition is to create a new public space – an urban living room that gives something back to the city. Dan Stubbergaard, architect and COBE's founder, explained that the design team want to create a place that provides optimal settings for the heroes of everyday life – the thousands of volunteers who make an extraordinary effort to help marginalized people.

The triangular building has an 850m² roof that acts as a large public staircase extending from street level to the second floor of the building. This unique stairway acts also as a terraced stand, making it an attractive and inviting meeting place for staff, volunteers and passers-by to use. It has certainly become a popular meeting place: in summer, volunteers and staff hold meetings here while locals use the steps for anything from workouts to coffee breaks.

COBE developed the project in close collaboration with the Red Cross and volunteer representatives. Everything from the spatial program to the design of the building was tested, discussed and adapted through dialogue, with great success. Since the building's opening in November 2017

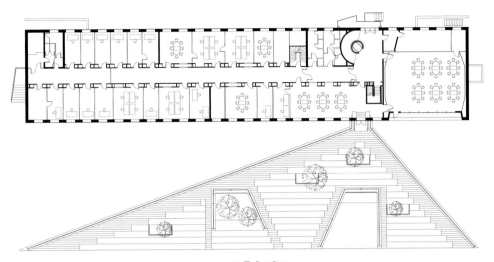
二层 first floor

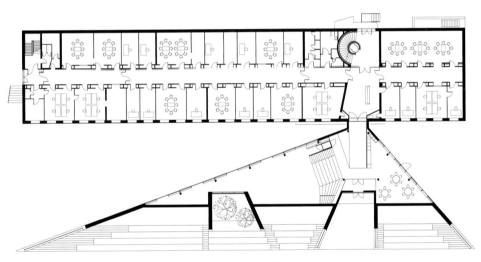
一层 ground floor

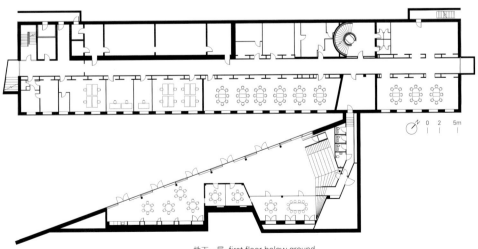
地下一层 first floor below ground

项目名称：Red Cross Volunteer House / 地点：27 Blegdamsvej, 2100 Copenhagen Ø, Denmark / 事务所：COBE / 工程师：Søren Jensen Rådgivende Ingeniørfirma 景观建筑师：PK3, COBE / 土耳其承包商：C.C. Brun / 客户：Danish Red Cross-The new building was made possible only with a grant of DKK 30.7 million from the private foundation A. P. Møller og Hustru Chastine Mc-Kinney Møllers Fond til almene Formål. / 用途：volunteer centre-exhibition spaces, project, meeting and conference facilities, training facilities, disaster management facilities and a café / 建筑面积：850m² / 总楼面面积：750m² / 竞赛设计时间：2013 / 竣工时间：2017 / 摄影师：©Rasmus Hjortshøj-COAST (courtesy of the architect)

the Red Cross has received increased visits from other local branches and countless curious visitors stopping by.
As Anders Ladekarl, Danish Red Cross secretary general, described, "This new building has an open and inviting architecture that is in keeping with our ambition for dialogue with the community."
The building's distinctive sloping roof simultaneously blends into and reinterprets the area's unique character, with the nearby Danish Freemasons' Hall, the adjacent park – Fælledparken – and the yellow-brick Red Cross headquarters, which was the former Copenhagen County hall. The Volunteer House's stepped roof is built using the same yellow bricks.
The extension, which has an interior floor space of 750m², sits partially below ground and serves as the main entrance to both the headquarters and the volunteer center.
To the left, the Volunteer House opens up into an auditorium with a capacity of over 100. This space leads to a series of conference rooms for training, meetings, events, presentations and film showings.
The stepped roof surface opens up towards the headquarters, which are visible from any position within the Volunteer House. The original building and the new extension are further linked by a green park. With this new common space for employees and volunteers to meet, ideal conditions have been provided for development and cooperation to flourish. In the words of Dan Stubbergaard: "The building has become an urban space and expresses both generosity and modesty while inviting the outside world in."

感知波浪

意大利第一个当代艺术中心于1988年诞生于普拉托。该中心由企业家Enrico Pecci构想,并为纪念他的儿子Luigi而捐赠给该市。该中心是在几个创始合作伙伴——公共机构和私人赞助人的支持下建造的。它的使命是通过临时展览、教学活动、表演和多媒体活动来增强对国内以及国外新兴艺术的敏感度。为了确保独特的艺术资产不会因为缺乏展览空间而被陈列在储藏室中,在21世纪初,Pecci中心决定将其展览空间扩大一倍,翻修由Italo Gamberini设计的原有建筑,因为其中一些地方已经变得很危险而且也已经过时了。同时,中心还要进一步扩展其文化项目。一旦工程完工,该建筑群将占地近10 000m²,包括新增的档案馆和收藏50 000多册的专业图书馆、拥有1000个座位的露天剧场、140个座位的电影院/礼堂、400个座位的演出空间、书店、酒吧/小酒馆和餐厅,以及工作坊和各种会议室。

受Pecci家族委托,建筑师Maurice Nio致力于研究中心和区域之间的相互渗透问题。他是总部位于鹿特丹的NIO建筑事务所的创始人。建筑师将原有建筑物全部保留下来并完全保持了原样,然后按照原来周围花园的轮廓建造了一座环形的新体量,这座新建筑面向公众开放。由于新的入口、书店和餐厅都位于通透的一层,因此该中心能够向城市开放,激发人们的好奇心,吸引人们争相在此活动,尤其是中心还有一个实验花园和一个作为缓冲区的大广场。整个建筑群的最高点是天线,这种天线一方面能够表达人们想要感受在这片土地上不断跳动的新的创造力的意愿,另一方面,它也揭示出,这是一个能激发想象力的场所,是一个重要的存在。

从早期的设计阶段开始,Maurice Nio就为这座新建筑选择了一个令人印象深刻的标题:感知波浪。这个标题暗示了它作为一个接收器(甚至是转换器)的功能,它将捕捉和传播当今的振动。在天线下面,新的功能空间和路线将原来的建筑与新的建筑连接起来。该项目基于对展览功能的详细修改建造,目前通过一个令人意想不到的、不寻常的物体向外界展示,该物体可以用许多不同的方式来解读。这种微妙而理性的语言似乎超越了当前将艺术中心视为大城市标志的国际趋势。通过对国际性景观的关注以及加强与地域概念的联系,Pecci中心今天正在扮演一个将持续几十年的重要角色。

Sensing the Waves

The first center for contemporary art in Italy was born in Prato in 1988. Conceived by entrepreneur Enrico Pecci and donated to the city in memory of his son Luigi, the center was built with the support of several founding partners – public institutions and private patrons. Its mission was to promote a sensitivity towards emerging art – both national

普拉托新 Pecci 当代艺术中心
New Pecci Center for Contemporary Art in Prato
NIO Architecten

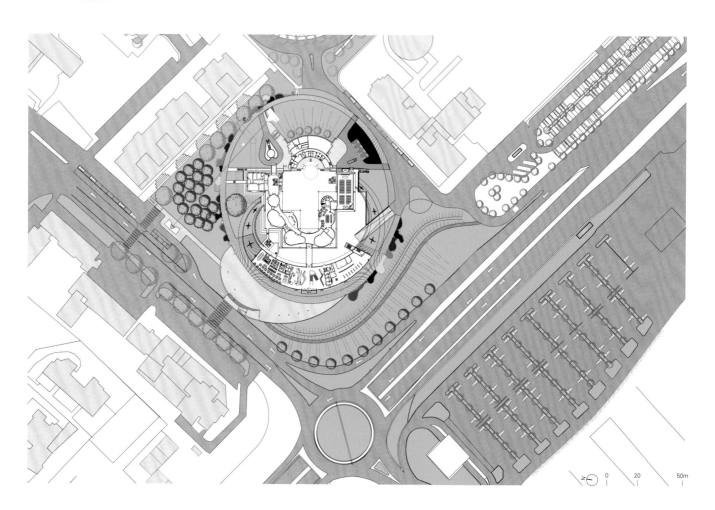

and international – by means of temporary exhibits, didactic activities, shows, and multi-media events. To ensure that unique art assets would not be sacrificed in the storerooms due to a lack of exhibition space, in the early 2000s, the Pecci Center decided to double its exhibition space and, at the same time, renovate the original building by Italo Gamberini, some aspects of which had become critical and obsolete, and further expand its cultural program. Once the works are completed, the complex will cover almost 10,000m², with additional archive and a specialist library holding over 50,000 volumes, an open-air theater with 1,000 seats, a cinema/auditorium with 140 seats, a performance space with 400 seats, a bookshop, a pub/bistro and a restaurant, in addition to workshops and various meeting rooms.

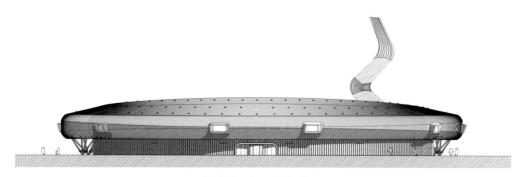

西南立面 south-west elevation

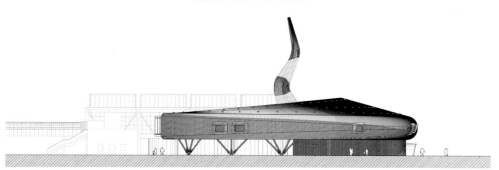

西北立面 north-west elevation

项目名称：New Pecci Center for Contemporary Art in Prato / 地点：Centro per l'Arte Contemporanea Luigi Pecci, viale della Repubblica 277, Prato, Italy / 事务所：Maurice Nio 指导：Luca Rimatori / 普拉托市政当局技术部门经理：Luca Piantini / 场地施工监理：Paolo Bartalini (2007~2012), Antonella Cacciato (2012), Massimo Lastrucci (2013~2016) Italo Gamberini所做的原建筑翻修的项目与施工监理：Antonio Silvestri / 景观建筑师：Luca Piantini, Michele Faranda / 地质学家：Deborah Bresci (2007~2011), Damiano Franzoni (2012~2016) / 结构工程师：Ingenieursbureau Zonneveld, Iacopo Ceramelli, Alberto Antonelli, Daniele Storai / 结构工作现场指导：Andrea Vignoli, Claudio Consorti 机械与消防系统：Dante Di Carlo / 电气工程师：Maurizio Mazzanti (CMA) / 场地安全协调：Paola Falaschi / 照明设计：Bernardo D'Ippolito (Kino Workshop) / 声学设计：Pietro Danesi / 静态测试：Massimo Perri / 行政管理：Stefania Galli / 总楼面面积：12,125m² (extension_7,815m²; original building_4,310m²) / 造价：EUR 14,400,000 (addition, landscaping and restoration of original building) / 设计时间：2006 / 竣工时间：2016 / 摄影师：©Fernando Guerra | FG+SG fotografia de arquitectura (courtesy of the architect)

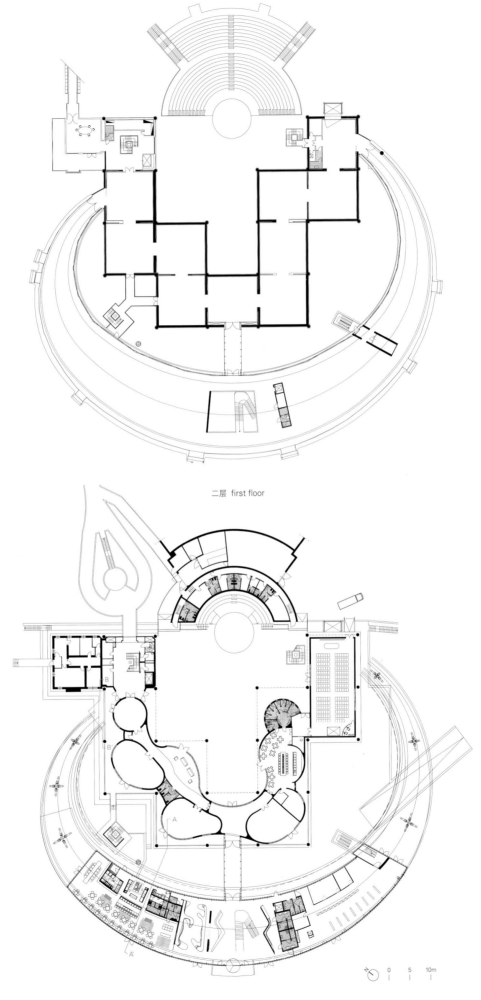

二层 first floor

一层 ground floor

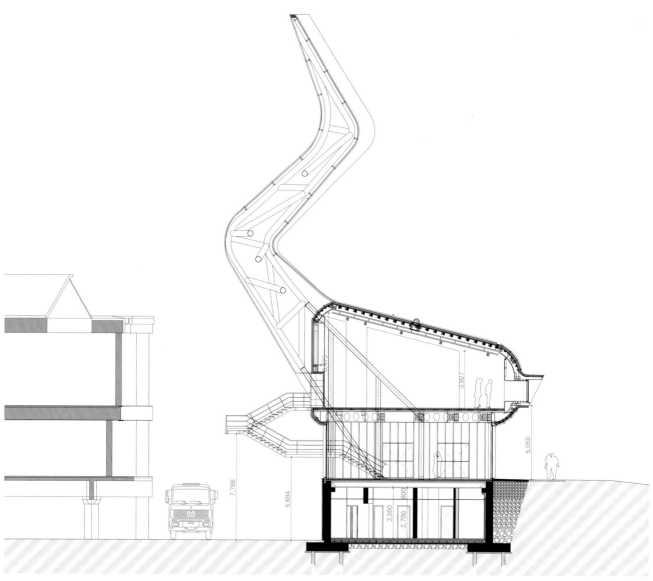

A-A'剖面图 section A-A'

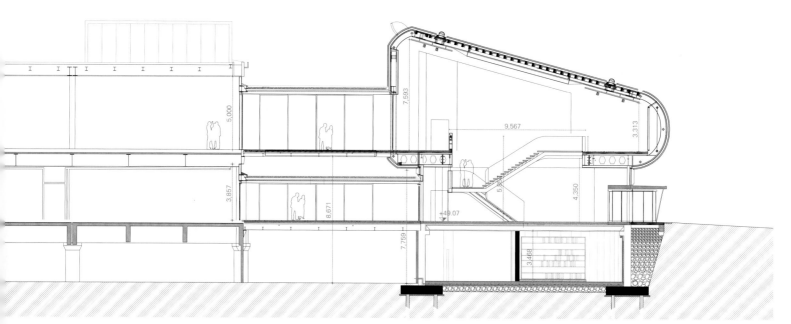

B-B'剖面图 section B-B'

Commissioned by the Pecci family, the architect Maurice Nio, founder of Rotterdam-based studio NIO Architecten, worked with the permeability between the center and its territory. The existing building is entirely maintained and left completely intact. A new volume in the shape of a ring, that traces over the contours of the original surrounding garden, is oriented towards the public dimension. Thanks to the new entry, the bookshop, and the restaurant, all located on the transparent ground floor, the center opens itself to the city, stimulating curiosity and inviting interaction, especially with an experimental garden and a large piazza as a buffer. The highest point of the whole complex is reached by a sort of antenna which is capable, on one side, of representing the will to sense the new forms of creativity which

are pulsing in the territory, on the other side, of revealing the important presence of a venue that will stimulate the imagination.

Ever since the early design phases, Maurice Nio chose a highly evocative title for the new building, Sensing the Waves, suggesting its function as a receptor (and perhaps even transmitter), that will capture and disseminate the vibrations of the present time. Underneath the antenna, a new map of functions and routes links the original building to the new building. The project is based upon a deliberate modification of the exhibition functions which are now revealed to the outside by the realization of an unexpected, unusual object that can be interpreted in many different ways. The subtle and reasoned language seems to go beyond the current international trend of realizing art centers as big urban icons. By confirming its attention towards the international scene and by strengthening its link to the territory, the Pecci Center is today undertaking a new important objective to be carried on for the next decades.

城市住宅新高度

New Heights
Urban

尽管住房危机正向全球的许多地方蔓延,并引起越来越多的来自政治层面的关注和辩论,但越来越多的住房项目正在我们身边兴建。还有很多人关注着这一发展,这些人有来自建筑领域的,也有来自环境方面的。因此建筑师和规划师在平衡各个层面的工作中扮演着重要的角色。这几乎是一项不可能完成的任务,毫无疑问,讨论将继续下去。世界人口的增长速度比以往任何时候都要快,现在世界上许多城市的土地资源和建设优质住房的预算资金都很短缺,而且所有人都需要一个舒适的住房。因此,建筑师、设计师和规划师受

Whilst the global housing situation is shifting into crisis in many parts of the world and catching ever more political headlines and debate, more and more housing projects are sprouting up around us. Extra eyes are watching this development from architectural and environmental aspects – so architects and planners have an important role to play in finding a balance that works on all levels. This is an almost impossible task, and the discussion will undoubtedly continue. The world's population is growing faster than ever before, and there is now, in too many cities, all over the world, a shortage of land and budgets for building good quality, and the necessary af-

北塔_Norra Tornen / Reinier de Graaf / OMA
79 & PARK公寓楼_79 & PARK / BIG
伦敦的储气罐_Gasholders London / WilkinsonEyre Architects
Novetredici住宅综合楼_Novetredici Residential Complex / Cino Zucchi Architetti
Loftwonen 61号楼_Loftwonen Block 61 / Architecten|en|en
城市住宅新高度_New Heights in Urban Housing / Heidi Saarinen

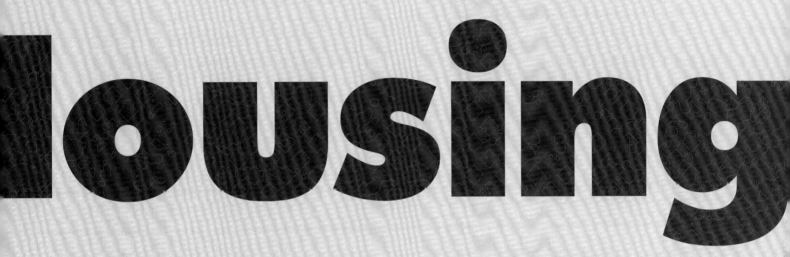

到了更为严格的审视，预期结果将是高水平的，为此可以在一切层面不惜一切代价。此外，在未来的许多年里，建造住房必须满足位置的要求，满足需要住房的人群、家庭和几代人在未来多年的需求。

此外，我们不要忘记不断上涨的住房成本，对于所有相关人员来说，这一点不容忽视，其中包括开发商、设计师、建筑师、土地所有者、地方当局、私人所有者或社会租户。因此，必须采取创新的、可持续的方法，我们将在本文中看到的项目就是通过不同的方式采用了这种方法。

fordable housing for all. Therefore, architects, designers and planners are even more scrutinised and results are expected to be high, at all costs and at all levels. Furthermore, housing has to be appropriate for the location, type of people, families and generations that require housing, for many years to come.
Moreover, let's not forget the rising cost of housing, for all involved, developers, designers, architects, landowners, local authorities, private owners or social tenants. Therefore an innovative, sustainable approach is a must, as we see, in different ways, in the projects featured in this essay.

城市住宅新高度
New Heights in Urban Housing

Heidi Saarinen

关于一些住房项目得到许可、建造和使用的速度，有许多对话和热烈的讨论。还有建筑转换的想法，就是为不同类型的使用者提供不同类型、不同质量的房屋。在英国，相关讨论（例如，在目前的一些建筑案例中）的关注点是私人住宅和社会住房问题，以及在混合用途住宅楼中设置"富"和"穷"两个独立入口的问题。在混合用途住宅楼中，私人住房和社会住房总是并存的。

那么，我们如何平衡不同的需求，解决私人和社会所有者/租户、场地和周围的便利设施以及景观之间的矛盾？在我们不断发展的城市和繁忙的生活中，我们如何才能充分利用健康的生活、社会互动、对城市环境的看法以及工作/生活/娱乐之间的整体平衡？我们如何在城区内创建积极的社区，而不是让人们居住在单独的无名公寓楼中？本文提到的项目都存在一些互动层面的理念，无论是居民之间互动，还是通过专门设计的通道、走廊和公共区域产生互动，或通过来自战略定位的公共花园、窗户、露台和阳台的视线互动。此外，还需要与城市本身建立更直接的联系。许多项目都位于城市基础设施旁，包括铁路、繁华的交通路段、零售区或使用频繁的道路和人行道。随着商业街区与在线零售商的竞争愈演愈烈，建筑师和规划师还必须设法找到零售市场和住房市场之间的互动方式，以全方位实现生产力的提高。毫无疑问，如今的建筑师们已经对所有方面进行了仔细的规划，包括位置、材料、构造、建筑高度和景观——在本文提到的项目的每个场地中，这些都是经过专门计算和调整的。

由BIG设计完成的79&PARK公寓楼，或称Kullen（178页）位于斯德哥尔摩Gärdet地区，是一个住宅项目。其奢华之处在于其紧邻一大片自然景观，但却靠近市中心。它的附近还有许多重要的地标建筑，包括博物馆和新野兽派的代表Kaknästornet建筑。[1] 79&PARK是一座大型的混合用途综合建筑，楼层不同，高度不同，户型面积不同，能满足不同居住者的需求。建筑师小心地将它设置在一个备受人们喜爱的自然场所之中，

There are many conversations and some heated discussions about the level of speed at which some housing projects are agreed, built and occupied. There is also the idea of diversion, different types and quality of housing for different types of occupiers. In the UK, the rhetoric is, in some current cases, for example, about private and social housing projects and the separate "rich" and "poor" entrances to mixed housing complexes where private and social occupancy occur alongside each other.

So how do we balance the different needs, for private and social owners/tenants, the sites and surrounding amenities and landscapes? How can we, in our ever-growing cities and within our busy lives, make best use of healthy living, social interactions, views into our city surroundings and overall balance of work/life/play? How can we create positive communities within urban areas rather than individuals living in separated, anonymous apartment blocks? The projects reviewed here, all have some aspect of interaction, whether it is amongst the residents, via pathways, corridors and communal areas deliberately designed into the projects, or via the views from strategically positioned communal gardens, windows, terraces and balconies. There is also a need for a more immediate connection to the city itself. Many of the projects are situated right next to the city's infrastructure, railways, a busy traffic section, retail areas or frequented paths and walkways. With High Streets competing with online retailers, architects and planners must also find ways to interact between the retail and housing markets, for all round productivity. All these aspects have no doubt been carefully planned by the architects, in terms of location, materials, construction, building heights and vistas – all calculated and adapted specifically for each of the sites in these featured projects.

79&PARK (p.178), or Kullen, on Gärdet in Stockholm, by BIG is a housing project that has the luxury of being situated right next to a vast area of natural landscape, yet close to the city center. There are also many important landmarks nearby,

79 & PARK公寓楼，瑞典
79 & PARK, Sweden

北塔，瑞典
Norra Tornen, Sweden

人们在那里可以散步、慢跑，并进行各种家庭活动和娱乐活动。斯德哥尔摩有几个精心设计的处在大自然中的住房开发项目，在这些地方，只要你站在门口，就可以感受到大自然的气息。这就是瑞典生活方式的核心，也几乎可以肯定地说是人们所期待的。

公寓，尤其是它们的位置是经过巧妙设计和安排的，以得到最佳的采光效果和视野。例如，公寓的东北角离公园最远，看向公园的视野不好，与公园联系也就没那么紧密，因此，东北角就被建得更高，以获得最佳的视野。公寓西南角延伸至Gärdet公园，这里被设计得较低，以使建筑与公园和自然紧密相连。出于方便和实用的考虑，所有的居住单元都是模块化、标准化和预制的。

坐落在斯德哥尔摩的新开发项目北塔（166页）包括两栋建筑，由OMA和Reinier de Graaf开发，它始于一个20年前的老项目，这个老项目最初曾被取消。历史情况是这样的：城市建筑师兼规划师亚历山大·沃洛达斯基设计了最初的提案——托尔塔，它被设计成一个在进入托尔斯潘地区和人口密集的哈加斯顿小区之前的一个可供"歇脚的地方"。

它是根据20世纪20年代斯德哥尔摩城市规划师提出的理念建造的，他们最初的设想是在这个地区建一座教堂，作为"社区的中心点"。但是教堂计划并未实施，取而代之的是托尔塔，它代表的是一座"当代象征主义建筑"。关于这项提议的许多争论都传到了社区和媒体上，争论的主要焦点是这座本应有着中低层住宅和较低城市景观的塔楼的高度。[2]

之后，设计人员对项目进行了重新设计，将原来"顶部更重"的设计进行"颠倒"，以达到最佳效果，并与原有场地实现更为流畅的联系。最终经过不同建筑师的多次努力，北塔这两座被命名为创新楼和螺旋楼（第二座塔楼于2019年竣工）的塔楼的设计权被OMA和Reinier de Graaf以及Oscar地产公司赢得。

including museums and the neo-brutalist Kaknästornet.[1] 79&PARK is a large mixed-use complex of dwellings with different levels, heights and different size, for a range of occupiers and their needs. It has been carefully placed on a popular natural spot where people walk and jog and carry out various family and recreational activities. Stockholm has several well-designed in-nature housing developments, where, if not immediately on the doorstep, nature is never too far away. It is central to the Swedish lifestyle, and is almost certainly expected.

The apartments and particularly their position for day lighting and views are cleverly designed and arranged. For example, the northeast corner, furthest from the park and therefore with less attractive views and immediacy to the park, has been built up higher to get the best views. The southwest part stretches out to the Gärdet park and has been designed for low-level living and for connecting closely to the park and nature beyond. For ease and practicality the units are modular, standardized and prefabricated.

An originally cancelled project, that started 20 years ago, was the beginning of this new development of two buildings, Norra Tornen (North Towers) (p.166) by OMA and Reinier de Graaf, also in Stockholm. The history is that Aleksander Wolodarski, a city architect and planner designed the original proposal, Tors Torn (Thor's towers), as a "place to pause", before entering Torsplan and the dense neighborhood of Hagastaden.

This was built on ideas by 1920's Stockholm city planners who originally envisaged a church in this area, as a "central point for the community". This never happened, and instead the towers would have been representing a "contemporary symbolic building". Much debate of this proposal took to the community and the media, with the main concern being the height of the towers in this otherwise low to medium level housing and cityscape.[2]

The development was then redesigned, and the original "top heavy" design was turned "upside down" to fit in better and allow a more fluid connection to the existing site. Finally, however, after several attempts by various architects to get the

伦敦的储气罐，英国
Gasholders London, UK

新设计在形式和描述上几乎呈现出科幻的色彩，结构由肋形彩色混凝土板组成，外侧嵌入了外露的多色石子骨料，体现出野兽派建筑的特征。建筑师创造了一系列独特的拼图式空间布局和巧妙利用日光照明的朝向（这一点对于冬天漫长而黑暗的瑞典非常重要）。窗户看起来像后加的盒子，在不同方向从立面向外延伸到城市中，从里面向外看就会发现，这些窗户都是全景窗户，不但能提供充足的日光，通过它人们还可以俯瞰整个城市和季节的变化。模块化的建筑方法使得许多组件的生产得以高效地进行，例如，整个浴室是在瑞典北部的工厂制造的，交付后在现场开槽安装，包括管道、瓷砖和板条也是如此。

预制混凝土板降低了成本，并在整个施工过程中都可以使用，因为在温度达到5°C后，混凝土就不能现场浇筑了。室内设计轻盈，并考虑到了通风的问题，设计和布局中带有现代主义的暗示。此外，居民还可以使用健身房、桑拿房、休闲室和电影院，还有一个客房公寓以及社交、创意活动空间。

位于伦敦国王十字街的三个前储气罐（194页）虽然建于19世纪80年代末，但它所展现出来的工程遗迹和技艺仍然为人们所钦佩，这要归功于WilkinsonEyre建筑师事务所的建筑师们。它于2000年停止使用，被列为二级列管建筑，如今经过翻新，被重新设计成一个具有开创性的住宅综合体，仍然能突出这一重要伦敦工业区的创新和历史。受人无比敬仰，在历史过程中地位重要的储气罐及其遗迹虽然在英国各地被拆除，但居民们和当地历史学家则孜孜不倦地试图拯救它们，他们积极投身于反对拆除这一重要工业建筑的决定的努力中。不幸的是，在许多情况下，这些努力都是徒劳的。具有讽刺意味的是，失败的原因是人们迫切地需要规划更多的住房。因此，国王十字街的储气罐建筑对整个城市及其工业遗产来说，是一个真正值得人们欢庆的工程。它的新拥有者们可以享受居住在摄政运河旁的生活，并从这些具有开创意义的新家中

commission, the two towers, Norra Tornen, named Innovationen and Helix (the second tower was completed in 2019), was won and designed by OMA and Reinier de Graaf with Oscar Properties.

The new design, almost sci-fi in form and description, consists of ribbed colored concrete slabs, embedded with exposed multi-colored aggregate pebbles on the exterior, echoing brutalist architecture. A range of unique spatially puzzle-like layouts and orientations with clever use of day lighting is created (much needed in Sweden during the long, dark winters). The windows appear as parasitic boxes extruding – stretching out over the city – from the facade in different directions. From inside they are panoramic windows, with lots of daylight and fantastic views over the city and it's changing seasons. Modular building methods allow for efficient manufacture of many of the components; for example, entire bathrooms are made in factories in the north of Sweden, to be delivered and then slotted into place on site, including plumbing, tiles and fittings.

Prefabricated concrete panels reduce costs and allow the construction to be carried on throughout the building process, as after 5 degrees Celsius concrete cannot be poured in situ. Interiors are light and airy, with a modernist hint in the design and arrangements. Additionally, the residents can enjoy a gym, a sauna, a relaxation room and cinema spaces. There are also a guest apartment and social and creative event spaces.

The engineering heritage and craftsmanship of the former Gasholder triplets in London's Kings Cross (p.194), built in the late 1880's can still be admired today thanks to WilkinsonEyre Architects. Having been decommissioned in 2000, the Grade II listed structures have now been refurbished and redesigned into a ground-breaking housing complex, still highlighting the innovation and history of this important London industrial site. Far too much admired and important historical gasholders and their heritage have already been demolished throughout the UK, whilst residents and local historians have tirelessly tried to save them, actively opposing the decisions to demolish this important industrial architecture.

Loftwonen 61号楼，荷兰
Loftwonen Block 61, the Netherlands

欣赏美丽的景色。建筑物的高度稍有不同，以反映过去这些美丽的维多利亚式建筑中气体水平的变化。

除了明显地反映出过去的历史特点，圆鼓的形式也被用来创造连接点。每栋建筑都有自己的核心筒和中庭空间，从那里，居民可以与环岛、环形人行道和用于进行社交互动活动的中央公共区域产生真正的和视觉上的联系。

储气罐框架经过漫长的仔细修复过程，终于修复完成，这项工作是由Shepley工程公司负责的。这些框架代表了一个重要的工业时代。为了确保其结构的完整性，中间经过了许多不同的精细的修复过程。在完成所有这些之后，它们被重新赋予了新的生命。"这个地区有一个真正的灵魂，它是伦敦的工业中心"，Shepley工程公司的修复主管特雷弗·马尔斯说。建筑师们说："铸铁柱头上有大写字母，相邻的横梁上有装饰性的标志，这实际上是维多利亚式建筑的一部分。"[3] 这突出了这一工业遗产的巨大重要性，以及这一时期工业建筑中材料、细节和装饰的重要性。在已完工的建筑中，原来的结构框架被钢和玻璃板包裹，为了遮阳和保护隐私，还安装了百叶窗，增加了立面的纹理和运动特质。

室内设计是开放式的，采用了不同的布置方式，还能欣赏到令人惊叹的景观。屋顶上有花园和社交空间，种植了绿植的屋顶吸引着大自然中野生动物的到来。

在荷兰，Architectenlenlen建筑事务所设计了公寓楼Loftwonen 61号楼（214页），位于埃因霍温的Strijp-S区。三座住宅楼的一部分受到周围城市景观及其工业历史上的形式和构造的启发。立面由红砖和预制混凝土砌筑成的住宅综合楼被仔细地融入原有的场地中。室内空间高大宽敞，楼层平面布局灵活，可以适应不同类型的生活需要。建筑与连接区之间的开放空间让我们想起了工业建筑风格。公寓走廊外有蓝色的工业钢楼梯和护栏。公共室内空间的统一配色方案与建筑正面的红砖立面形成了鲜明对比，同时仍然强烈暗示了周边的环境和过去的工业影

Sadly in many cases, this has been unsuccessful due to, ironically, planning and urgencies for further housing. Therefore, the Kings Cross gasholders are a true joy and celebration to the city and its industrial heritage. Its new occupiers can enjoy life by the Regents Canal and look out over beautiful views from these pioneering new homes. The heights of the buildings have been set at slightly different levels, to echo the movement of the gas levels when in operation in the past life of these beautiful Victorian structures.
Alongside the obvious historical past, the forms of the circular drums have been used to create connecting points. Each building has it's own core and atrium spaces from where the residents can physically and visually connect to the surroundings and the circular walkways and central common areas for social interactions.
The gasholder frames are carefully restored through a long and detailed process by Shepley Engineers. The frames represent an important age of industry, and they are lovingly restored back to life, through many different and intricate processes, making sure structural integrity is ensured. "The area has a real soul, it is London's industrial heart", says Trevor Marrs, Head of Restoration at Shepley Engineers. "The cast iron columns have capitals on them, the adjoining beams have decorative emblems and this was really just part of Victorian architecture of the day", say the architects.[3] This highlights the immense importance of this industrial heritage and the significance of material, detail and ornament in industrial architecture of this time. In the completed buildings, the original structures have been cladded in steel and glass panels, with shutters, for shade and privacy, adding texture and movement to the facade.
Interiors are open and arranged in different configurations with amazing views. There are gardens and social spaces on the roof, with green roofs that attract wildlife.
In the Netherlands, Architectenlenlen designed apartment building Loftwonen, Block 61 (p.214), in the Strijp-S district of Eindhoven, part of three housing blocks was inspired by the surrounding city landscape and also its industrial past

响力。正如本文中提及的其他项目一样，住在这里，无论是有意还是偶然，你都会有大把的机会结识你的邻居，与他们进行社交，这一点始终是所有公共城市生活的积极因素所在。

在米兰，Cino Zucch Architetti建筑事务所设计了Novetredici住宅综合楼（206页）。该综合楼位于新老区域之间，北面是城区，南面是原有的较传统的区域。建筑呈现出一种折中的、错列的外观特征，仿佛邀请我们进入一个空间迷宫，并引导我们走入分布在两座建筑之内的公寓。这两座建筑中较低的一座在西侧，较高的一座在东侧，二者通过一层的玻璃公共入口大厅连接。建筑师细致考虑了周边的城区，以便在新建筑与原有场地之间搭建起一种联系，尤其是科莫大街的历史建筑和巴勒莫新门新开发项目。双层高的公寓位于顶层，视野开阔，而且室内通风良好。

实心墙与大露台和大小不同的阳台混合在一起，给人一种"居住屏风"的感觉，这恰恰是建筑师的说法。人们可以瞥见公寓内的日常生活：在阳台上享受美好的时光，在一层台阶上与邻居交谈，或者在公共入口大厅收自己的邮件。居民可以享受私人花园带来的快乐，这是该开发项目的一个关键特征，居民还可以使用地下停车场。

建筑师从材料、颜色、纹理和资源使用等方面进行细节方面的整体设计，使方案具有独特性。同时，不同尺寸的开口、窗户、门和阳台使得设计充满变化，并满足了不同用户的需求。在建筑外部，被漆成灰绿松石金属色调的波纹穿孔铝板，以及不同类型的手工砖，塑造了这座城市住宅综合楼的统一风格。

in form and construction. With a facade of red brickwork and prefabricated concrete, this residential complex has been carefully slotted into the existing site. Interiors are tall spaces with flexible floor plans, for different variations of living. Open spaces between the buildings and connecting zones, remind us of industrial buildings with sharp blue industrial steel staircases and guardrails outside the apartment corridors. This uniform color scheme to the communal interiors creates a real contrast to the front redbrick facade, whilst still hinting strongly to the scheme's surroundings and past industry influences. As with the other projects covered in this essay, there are opportunities here to meet neighbors and have social exchanges, whether intended or by chance, always a strong positive factor of any communal city living.
In Milan, Cino Zucchi Architetti have designed the Residential Complex Novetredici (p.206). The complex sits in between the new and the old, urban areas to the north and the existing more traditional locations to the south. An eclectic staggered exterior invites us into a maze of spaces and leads us to the apartments, organised over two buildings: a lower building on the west side and a taller one on the east – joined by a communal glazed entrance hall on the ground level. The surrounding urban areas have been considered carefully, in order to allow connections between the new building and its existing locale, particularly the historical buildings of Corso Como and the new Porta Nuova developments. Double height apartments can be found on the top floors, with fabulous views and airy interior spaces.
Solid walls mixed with large terraces and balconies of different sizes give the impression of an "inhabited screen", according to the architects. Glimpses of everyday life can be seen, people enjoying time on their balconies, talking to neighbors on ground level in the terraced areas or collecting their mail from the common entrance hall. Residents can enjoy a private garden, a key feature of the development, and make use of the underground car park.
Holistic detailing in materials, colors, textures and use of resources, give the scheme identity, whilst the varied sizes of openings, windows, doors and balconies allow changeability within the design, and diverse user needs. On the exterior, there are corrugated and perforated aluminum sheets enamelled in a gray-turquoise metallic hue, and different types of

1. Kaknästornet.se (n.d.) Kaknästornet: Our History [Online]. Available at: https://www.kaknastornet.se/our-history/ [Accessed 09 Jan 2019].
2. Epstein, M. (n.d) Stockholmdirect: Mina torn var betydligt resligare [Online]. Available at: https://www.stockholmdirekt.se/nyheter/mina-torn-var-betydligt-resligare/repqam!rpnki8VXvv3pIL61V9mJ2A/ [Accessed 12/01/19].
3. Gasholderslondon.co.uk (n.d) Gasholders London: Architecture [Online]. Available at: https://gasholderslondon.co.uk/architecture
4. Sargent, E. (2018) Living with Buildings: Health and Architecture [Exhibition publication]. Exhibited at the Wellcome Collection, London 4 October 2018 – 3 March 2019

在城市区域中，选址、朝向和密度这些因素一直是住宅开发项目的关键。这里介绍的方案在某些方面有相似之处。在大多数城市地区，特别是在中心地区，都缺少有趣、适当和适用的场地。因此，新的城市住宅项目往往都是些密度很高的高层建筑——这正如我们在本文提及的许多项目中看到的那样。我们希望能通过建筑材料的使用，在外观、形式和表皮上增加个性。分层次地组织和规划观景视野，居民就可以从厨房、卧室和其他生活空间看到外部的景观，这也是一个设计关键点。我们将公共区域的出入口和动线视为一个重要因素，以促进居民之间的偶然会面和人与人之间的互动。住宅开发计划中经常会增加健身房和电影院等附加服务，这些也是吸引人们参与"城市生活梦想"的另一种方式。

埃米莉·萨金特，伦敦维康收藏博物馆居于建筑分馆馆长，将这一点完美地展现在人们的视野中，"建筑师、规划师和设计师的方法对个人的健康和社区的福祉将产生并已经产生了巨大的影响，同时也反映出身处社会之中，我们更应该优先考虑的广泛的事项。"

住房一直是建筑和社会经济辩论的关键话题。毕竟，住房和家庭是人们度过他们大部分高质量自由时间的地方，在家里可以与家人在一起、活动、玩耍和休息。有趣的是，正如我们所知，城市规划师对我们所应该秉持的生活方式，对我们的住房应该如何融入（或不融入）我们的城区有很多想法。更高的房屋购买价格和租赁成本迫使许多人，特别是年轻人，离开城市中心区域，并迫使他们为了更便宜的生活成本而进一步远离。我们对未来充满兴趣，随着我们继续在城市和城区建造住宅和混合用途开发项目，我们希望，它们将成为供所有人使用的持久、健康和令人兴奋的场所。

handmade bricks, all unifying the style of this urban housing complex.
Site, orientation and density of the urban location always play a key role in housing developments. There is a similarity between some of the aspects in the schemes featured here. In most urban locations, particularly in central settings, there is a shortage of interesting, appropriate and available sites. Therefore new urban housing projects tend to be dense, tall, high-rise – with, as we have seen here in many of the projects, a desire to add personality to the exterior, the form, the skin, through materiality. Organising and planning hierarchically the view, the residents should see from their kitchens, bedrooms and other living spaces, and this is also a key design move. We see access and circulation in communal areas as an important element, allowing for chance meetings and human interactions between the residents. Additional services such as gyms and cinema rooms are often added to the housing schemes, another way to attract people to take part in the "city living dream".
Emily Sargent, Curator of Living with Buildings at the Wellcome Collection, London, puts this perfectly into perspective, "The approach of architects, planners and designers have – and have had – a huge influence on the health and wellbeing of individuals and communities, while also reflecting wider priorities in society".[4]
Housing has always been a key player in architecture and socio-economic debate. After all, housing, the home, is where people spend most of their high-quality free time, with family, activities, play and rest. Interestingly, urban planners have had, as we know, many ideas about the way we should live, how housing should fit in (or not) into our urban areas. Higher costs of buying and renting push many people, particularly young people, out of the central zones, and force them to move further out for cheaper living costs. We look with interest at the future and as we continue to build homes and mixed-use developments in our cities and urban areas, we hope that they will be long lasting, healthy and exciting places for all.

北塔
Norra Tornen

Reinier de Graaf / OMA

北塔项目开始于两栋建筑遗留的外围护结构设计方案,这两栋建筑最初由斯德哥尔摩前城市建筑师Aleksander Wolodarski设计,但设计还未完成项目就被取消。大楼不是以一块平板或者塔楼的形式出现的,而是由高度逐渐增高的不同部分组成,这样的设计不会在城市中显得过于突兀。而最终的设计方案则让每间公寓都拥有开阔的户外空间,避免了将对外围护结构的理解照搬到建筑形式之中。

根据弗洛伊德的"向前飞行"理论——为了战胜最初的恐惧,我们要热情地拥抱不可避免之物——在这个项目中,建筑师必须接受之前规划好的建筑外围护结构,并在此基础上做出突破。他们利用横向分割对大楼原本单调的纵向分割进行了补充,赋予建筑外观一种单一的、同质的处理方式,创造了一系列交替设置的凹进户外空间和凸出起居室,形成丰富的粗糙表面。建筑材料选用肋形彩色混凝土,混以不同颜色的石子骨料,体现出野兽派建筑的特征。据建筑评论家Reyner Banham说,野兽派建筑这一说法是由瑞典建筑师Gunnar Asplund的儿子Hans Asplund提出的,他在一封寄给英国建筑师友人的信件中描述自己同事的作品时第一次使用了这一说法。

北塔项目采用预制混凝土板建造,这项施工技术使得工程可以在5℃以下——这种温度下现浇混凝土就无法流动了——的环境中继续进行。此外,预制混凝土还大幅度削减了工程造价,使得墙地比接近1的设计方案成为可能。而在一般的项目中,0.5的墙地比所需的造价就会让开发商望而却步。由此节省下来的经费可以被用来创造多朝向、大开窗的独特户型,这对于瑞典这个一年中有一半时间都缺乏阳光照射的国家来说尤为珍贵。

北塔项目位于市中心,周围的建筑大部分都是在第二次世界大战之前建造的住宅群,为斯德哥尔摩带来了新的生活方式,人们可以在高密度的生活中享受属于自己的户外空间。别忘了,斯德哥尔摩的空气质量在所有欧盟国家的城市中可是高居第四位。

两座塔楼中的第一座——创新楼中包含182间公寓,从44m²的一室住宅到271m²的顶层公寓。大部分住宅为两室或三室,面积在80m²到120m²之间。除了居住,建筑中还设有一个放映室,一个聚会用餐厅,一间客房公寓,一个配有桑拿房和休息室的健身房,以及一层配套零售空间。旁边的螺旋楼包含138间公寓以及便利设施。

两栋建筑的高度分别为125m和110m,是斯德哥尔摩市中心最高的住宅楼。它们位于斯德哥尔摩北部的Hagastaden区域,这里是一个围绕Karolisnka学院(诺贝尔医学奖颁奖机构)发展起来的新区,两栋建筑正是这座城市的门户。建筑师将原来规划的纪念性建筑物的外围护结构彻底转变成了以家庭生活为重的建筑。多种多样的公寓让曾经死板的结构呈现出惊人的活力,甚至可以说是展现出了人性的光辉。

斯德哥尔摩，1868年
Stockholm, 1868

北塔印象，从西面观看的视野
impression of Norra Tornen, seen from west

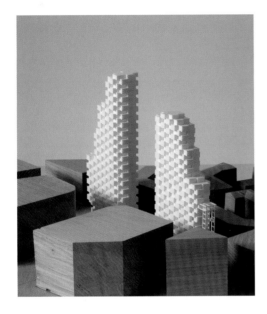

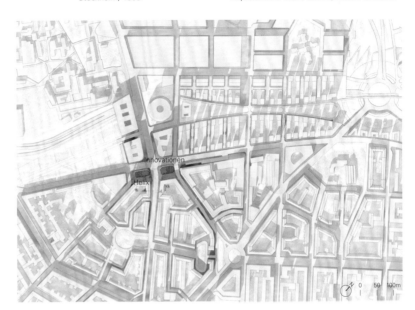

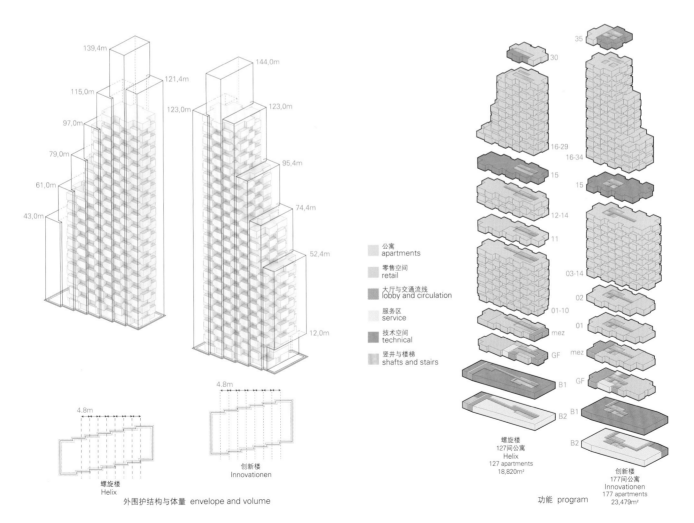

外围护结构与体量 envelope and volume

功能 program

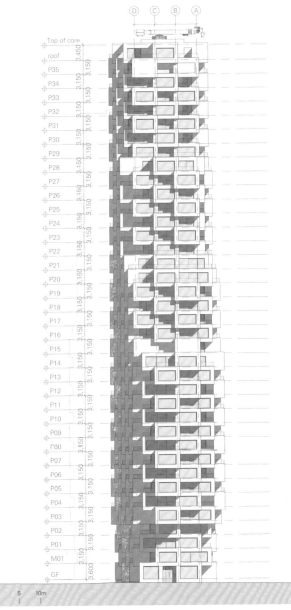

东立面——创新楼
east elevation_Innovationen

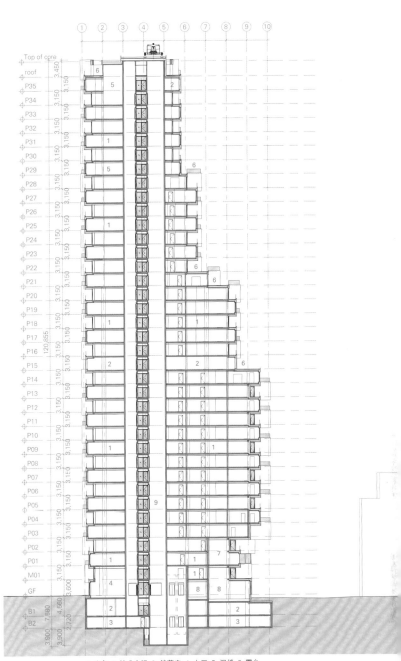

1. 公寓 2. 技术空间 3. 储藏室 4. 大厅 5. 阁楼 6. 露台
7. 便利设施 8. 零售空间 9. 电梯井
1. apartment 2. technical 3. storage 4. lobby 5. penthouse
6. terrace 7. amenity 8. retail 9. lift pit

A-A' 剖面图——创新楼
section A-A'_Innovationen

The Norra Tornen (North Towers) project starts with two inherited building envelopes, the remains of a canceled project initiated by the former Stockholm city architect Aleksander Wolodarski. Each kind of "crescendo" composition of different heights – neither slab nor tower – prohibit the unfolding of an uncompromised typology. Conversely, the opted program – apartments with an emphasis on large outdoor spaces – prevents too literal a translation of the envelopes into architectural form.

Through a kind of "Freudian flight forward" – a passionate embrace of the inevitable in order to conquer and overcome one's initial fears – the prescribed building envelope is adopted as a given one. Its initial vertical segmentation is complemented by a second, horizontal segmentation that gives the buildings' exterior a single, homogeneous treatment: a rough skin, formed by an alternating pattern of withdrawn outdoor spaces and protruding living rooms. The chosen material, ribbed colored concrete brushed with exposed multi-colored aggregate pebbles, deliberately echoes brutalist architecture. According to critic Reyner Banham, the term "brutalist" was coined by Swedish architect Hans Asplund – son of Gunnar – when referring to a design by studio colleagues, in a letter to British architect friends.

Norra Tornen is built from prefabricated concrete panels – a construction technique that allows work on the building site to continue in temperatures below five degrees Celsius, the limit which prohibits in situ concrete pouring.

Prefabrication also significantly reduces construction costs. A design with a wall-to-floor ratio close to 1 (most developers are discouraged by even a 0.5 ratio) is suddenly not an unthinkable proposition. The investment can be channeled into creating apartments with unique layouts, multiple orientations and larger windows – a precious asset in a country with scarce daylight for half of a year.

In a city center with a housing stock largely built before the Second World War, Norra Tornen introduces a new way of living. It brings together density with the possibility to enjoy

level 30, 32 _ Innovationen

level 29 – penthouse with large roof terrace _ Innovationen

Innovationen tower apartments

- 2 room apartment
- 3 room apartment
- 4 room apartment
- 5 room apartment
- public/service
- technical

level 5, 7, 9, 11, 13 _ Helix

level 5, 7, 9, 11, 13 _ Innovationen

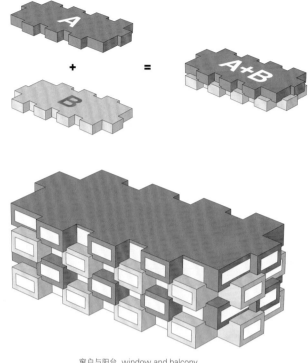

窗户与阳台 window and balcony

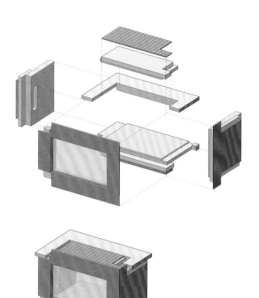

凸窗 bay window

outdoor space: Stockholm already ranks fourth among EU cities with the highest air quality.

The first of the two towers – The Innovationen Tower – comprises 182 units, ranging from 44m² one-bedroom apartments to a 271m² penthouse on the top floor, with the majority consisting of two- or three-bedroom apartments of 80m² to 120m². The residential units are complemented by a cinema room, a dining room for parties and celebrations, a guest apartment, a gym with a sauna and a relaxation area, and retail spaces on the ground floor. The neighboring Helix Tower includes 138 units, plus amenities.

At the heights of 125m, and 110m, respectively, the towers are the highest residential buildings in central Stockholm. Located in Hagastaden, a new district in the north of Stockholm developed around the Karolisnka Institute (which awards the Nobel Prize in Medicine), they stand as a gate to the city. However, the manipulation of the initial building envelopes radically transforms their initial implied architecture of monumentality, by giving way to an articulation of domesticity. A once formalist structure now houses apartments that are surprisingly informal…one could even say humanist.

项目名称：Norra Tornen / 地点：Torsplan 8, 113 65 Stockholm, Sweden / 事务所：OMA / 合伙人负责人：Reinier de Graaf / 结构、机械工程师：Arup 立面工程师：Arup Façade Engineering / 当地工程师：Sweco / 消防安全：Tyréns AB / 声学设计：ACAD / 法规咨询：Tengbom / 客户：Oscar Properties 功能：320 apartments (Helix-138, Innovationen-182), retail, sercives / 用地面积：Helix- 575m²; Innovationen-660m² / 建筑面积：Helix-14,039m²; Innovationen-17,787m² 总楼面面积：Helix-18,820m²; Innovation-23,479m² / 高度：Helix-110m (32 floors); Innovationen-125m (36 floors) / 材料：colored concrete ribbed facade, brushed with an exposed multi-colored aggregate pebble mix / 设计竞赛时间：2013 施工时间：Innovationen-2015.12~2018.12 / 交付使用时间：Innovationen-2018.9; Helix-beginning 2020 / 摄影师：©Laurian Ghinitoiu (courtesy of the architect) (except as noted)

79 & PARK 公寓楼
79 & PARK
BIG

79&PARK公寓楼 79&PARK
场地被诠释为周围公园的延伸，二者处在同一个城市街区中
The site is interpreted as an extension of the surrounding park, all within an urban block.

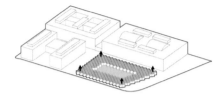

挤压 Extrusion
一个规则的方形网格经过挤压，创造了一座环形建筑，围绕场地边缘设置
A regular grid of squares is extruded to create a perimeter block that borders the site.

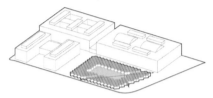

多孔 Porosity
三条公共通道活跃了作为中央共享空间的庭院
Three public passages activate the courtyard as a shared, central space.

日光 Daylight
建筑西南角降低，使庭院内可以直接接受日光照射
The south-west corner is lowered to provide direct sunlight to the courtyard.

地标 Landmark
西北角抬高，创造了一个城市地标
The north-east corner is raised to create an urban landmark.

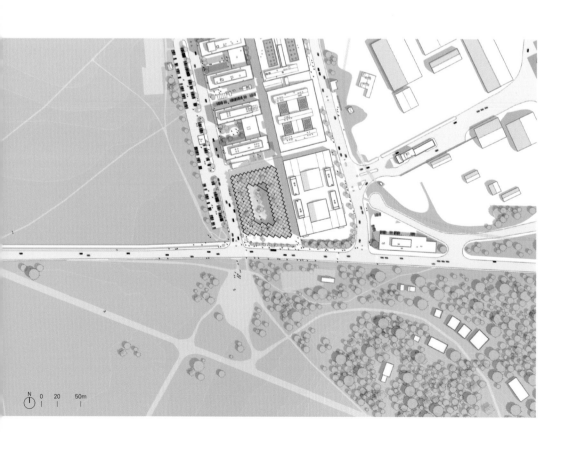

丹麦建筑事务所BIG, Bjarke Ingels Group, 在斯德哥尔摩完成了一个名为79＆PARK的模块化公寓楼项目。该项目是建筑事务所与Oscar地产公司合作开发的, 旨在扩建位于其前面的皇家国家城市公园。79＆PARK的设计类似于一个枝叶繁茂的木质山坡, 在其立方体结构的顶部有一片绿色梯田的设计。

设计团队有意识地决定在植被丰富的国家公园Gärdet的边缘, 提供一个敏感度高并能够让人肃然起敬的建筑形式, 同时还能让住宅中的住户都能欣赏到壮观的景色。为了直接对周围环境做出回应, 建筑师将西北角和东南角设计得与邻近的建筑同样高; 而东北角离公园最远, 理论上视野最差, 因此建筑师在设计上将这个地方拉起, 使其成为公园和港口最壮观的景色。

建筑的西南角最远延伸至Gärdet公园, 并在建筑和自然之间创造了一个具有人文气息的边缘联系。建筑师通过将其向下推至能够延伸的最低的位置, 将它转变为一个公共平台, 这里可以270°看到公园景观, 同时大多数住宅单元都可以欣赏到公园的美景。

同样的设计特点也确保了中央庭院总是能得到充足的阳光。为了进一步体现出对Gärdet国家公园的崇敬之意, 该建筑的大部分区域通过像素一样的设计在视觉上降低了体量的大小, 并按人体尺度进行缩放。这样的设计不仅使建筑的表现形式更加有组织, 完美地反映出周围的景观, 而且还提供了一种通过使用标准尺寸的预制单元以可控且廉价的方式顺应地形建造的方法。79＆PARK被描述为"向天空开放的绿洲"。建筑一层提供一系列服务, 包括一所幼儿园。169户公寓的位置以及相对于邻居的位置都经过仔细设计, 避免如同像素化般的公寓楼内不同住户之间产生直接的视觉接触, 以此来保护他们的隐私和自

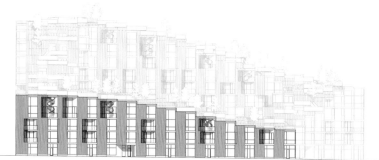

西立面 west elevation

南立面 south elevation

由。

方形模块的尺寸为3.6m×3.6m，采用阶梯式设计，以创造最高点达35m的建筑。公寓楼的公共设施包括一家狗狗日托中心、自行车停车场和一所幼儿园。

BIG创始合伙人Bjarke Ingels说："79&PARK是一个逐级降低的居住景观，将郊区住宅的壮丽景色与城市生活的品质联系起来。"

"城市中心绿洲化的社区带来的亲密关系创造了一种和平与安宁的感觉，同时也让居民感觉到自己属于79&PARK大社区的一部分。从远处眺望，79&PARK就像斯德哥尔摩市中心的一座人造山坡在那里屹立着，"他补充道。

Danish architecture practice BIG, Bjarke Ingels Group, has completed a modular apartment block in Stockholm entitled 79&PARK. The project has been developed in collaboration with Oscar Properties, and is intended as an extension of the royal national city park that lays in front of it. The design of 79&PARK resembles a verdant wooden hillside, incorporating a succession of green terraces on top of its cubic structure.

Located on the edge of Gärdet, a treasured national park, the design team made the conscious decision to provide a sensitive, respectful form while allowing the same choices to simultaneously manifest as exceptional residences with spectacular views. In direct response to its context, the corners in the north-west and south-east take the heights of their immediate neighbor, while the north east corner, furthest from the park and nominally with the worst view, is pulled upwards to grant it the most spectacular views of both park and port.

The south-west point of the building extends the furthest into Gärdet, and creates a humane edge between building and nature. This is pushed down to the lowest profile, transforming it into a public platform with a 270-degree view of the park landscape and simultaneously allowing the majority of the residential units to also enjoy views of the

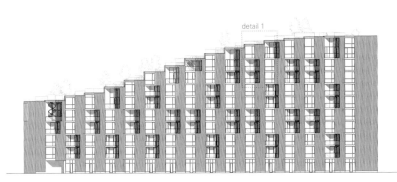

东立面 east elevation

北立面 north elevation

park.

The same design feature also ensures that the central courtyard will always receive copious amounts of sunlight. In further deference to the Gärdet national park, the mass of the building is visually reduced through a language of pixels, scaled to the human form. This manipulation not only allows for a more organic expression, perfectly reflecting the surrounding landscape, but also provides a way to accomplish the building topography in a controlled and inexpensive way through the use of prefabricated units of standardized sizes.

Described as "an oasis opened towards the sky", 79&PARK features a series of services on the ground floor, including a kindergarten. The positioning of each of the 169 apartments, vis-à-vis their neighbors, is designed to avoid direct lines of visual contact between the inhabitants of the pixelated apartment block, safeguarding their privacy and autonomy. Square modules, measuring 3.6m by 3.6m, are arranged in a stepped design to create a building that reaches to 35m at its highest point. Communal facilities for the apartment block include a dog daycare center, bicycle parking, and a preschool.

"79&PARK is conceived as an inhabitable landscape of cascading residences that combine the splendours of a suburban home with the qualities of urban living," said Bjarke Ingels, founding partner of BIG studio.

"The communal intimacy of the central urban oasis offers peace and tranquility, while also giving the residents a feeling of belonging in the larger community of 79&PARK. Seen from a distance, 79&PARK appears like a manmade hillside in the center of Stockholm," he added.

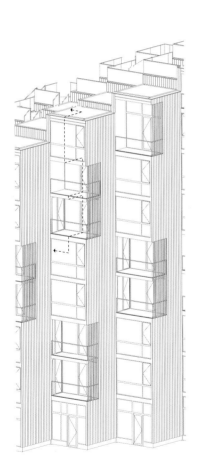

详图1 detail 1

a-a' 剖面图 detail a-a'

项目名称：79&PARK / 地点：Stockholm, Sweden / 事务所：BIG / 合伙人负责人：Bjarke Ingels, Jakob Lange, Finn Nørkjær / 项目经理：Per Bo Madsen / 项目负责人：Cat Huang, Enea Michelesio / 项目建筑师：Høgni Laksáfoss / 施工建筑师：Jakob Andreassen, Tobias Hjortdal, Henrik Kania / 项目团队：Agata Wozniczka, Agne Tamasauskaite, Alberto Herzog, Borko Nikolic, Christin Svensson, Claudio Moretti, Dominic Black, Eva Seo-Andersen, Frederik Wegener, Gabrielle Nadeau, Jacob Lykkefold Aaen, Jaime Peiro Suso, Jan Magasanik, Jesper Boye Andersen, Jonas Aarsø Larsen, Julian Andres Ocampo Salazar, Karl Johan Nyqvist, Karol Bogdan Borkowski, Katarina Macková, Katrine Juul, Kristoffer Negendahl, Lucian Racovitan, Maria Teresa Fernandez Rojo, Max Gabriel Pinto, Min Ter Lim, Narisara Ladawal Schröder, Romea Muryn, Ryohei Koike, Sergiu Calacean, Song He, Taylor McNally-Anderson, Terrence Chew, Thomas Sebastian Krall, Tiago Sá, Tobias Vallø Sørensen, Tore Banke / 合作者：Acad International, Andersson Jönsson Landskapsarkitekter, BIG IDEAS, De Brand Sverige, Dry-IT, HJR Projekt-El, Konkret, Metator, Projit, Tengbom, HB Trapper / 客户：Oscar Properties / 用途：Housing / 总楼面面积：25,000m² / 竣工时间：2018 / 摄影师：©Laurian-Ghinitoiu (courtesy of the architect)

四层 third floor

一层 ground floor

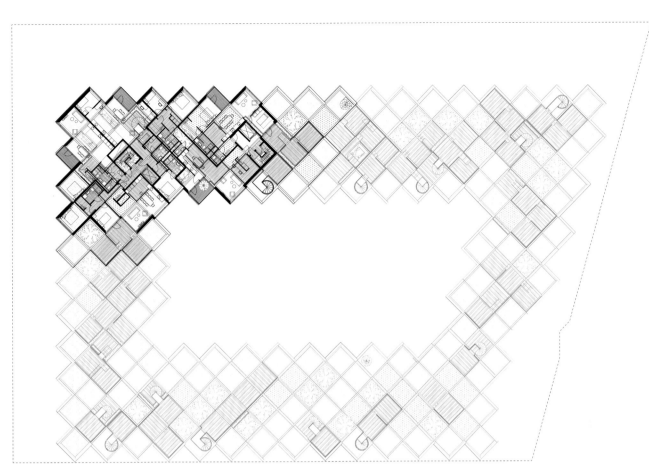
十层 ninth floor

七层 sixth floor

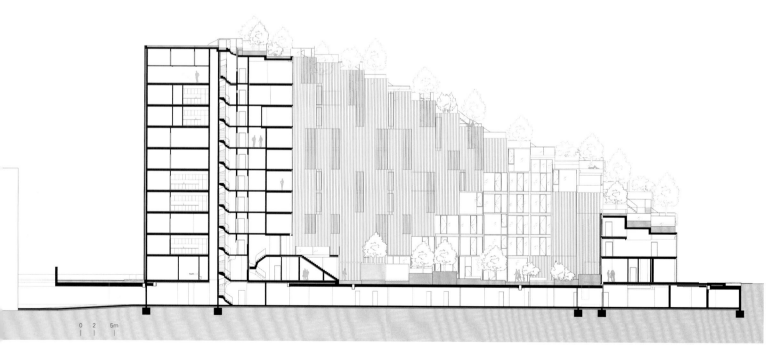

A-A' 剖面图 section A-A'

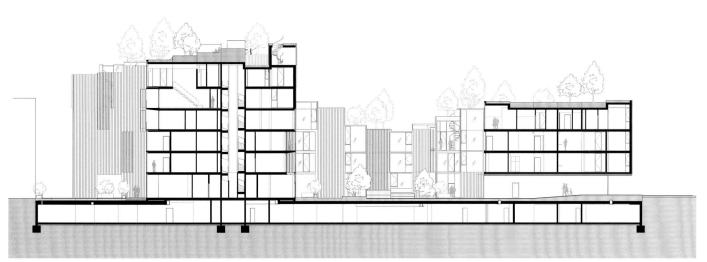

B-B' 剖面图 section B-B'

伦敦的储气罐
Gasholders London
WilkinsonEyre Architects

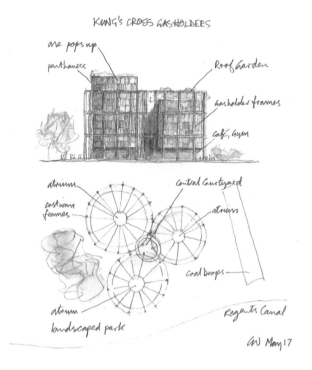

2018年，WilkinsonEyre建筑师事务所完成了对伦敦储气罐的设计工作，该事务所在三套列管储气罐指南框架内共开发了145套公寓。

国王十字街项目是欧洲最大的城市重建计划，其丰富的工业遗产与它的复兴密不可分。其中最有特色的保留项目是属于二级列管建筑的三套储气罐项目，以及建于1867年的铸铁和气柜导轨框架。由于重工业迁至城市郊区，这三套建筑被遗弃，在2001年，由于考虑到隧道铁路的连接问题而被拆除。

包括123根柱子在内的导轨框架经过了精心的修复，尽管它们已经有150多年的历史了，但它们目前的状况非常好，这主要是因为经过了32层防腐油漆的保护处理。

据创始董事克里斯·威尔金森称，该设计的意图是保留结构，但同时赋予它新的含义，并在未来的用途方面加以考虑。WilkinsonEyre的获奖设计的理念是将三座住宅建筑安置在优雅的框架内。设计中提出建立三个不同高度的容纳空间，以此映射出原先储气罐根据内部气体的压力而上升或下降的移动感觉。第四个中心虚拟空间形状在铸铁结构的交叉处形成一个开放的庭院。储气罐的设计是为了创造新旧之间的动态对比。沉重的工业美学和原始的物质性与室内空间设计所体现出来的轻盈和精细形成鲜明对比。

该项目一共提供145套公寓、一个私人健身房和水疗中心、一个商务休息室和一个娱乐会所。公寓部分通过中央庭院进入，每个鼓形的

体量都有自己的中庭和核心筒。它们都由一系列环绕庭院的圆形通道连接,位于中央的水景会反射光线。在另一个具有反差感的设计中,屋顶被设计成花园,为这个重新被用作居住功能的城市景观带来了自然的氛围。

正如威尔金森所说:"几何圆形的使用产生了非常美妙的想法。最初这仅仅是一个挑战,结果却成了一种福祉。"建筑的圆形特征使得公寓可以进行自然采光,同时起居室和卧室被布置在周边。这种"馅饼状"结构形成了开放式的公寓结构,视野开阔,朝向多样。

建筑覆层由模块化、垂直的钢和玻璃面板组成,并使用电子控制百叶窗来进行遮挡,从而为居住者提供荫蔽,保护他们的隐私。该公寓专为繁忙的城市居民设计,采用创新技术,使居民能够体会、控制以及在环境中调整他们的生活空间。

对储气罐的新构想满足了人们对现代生活的功能要求,同时也是对三重导轨框架的特点和形式的一种点赞。该项目创造性地响应了工业遗产的需求,以富有想象力的设计解决了固有的限制问题。

In 2018, WilkinsonEyre completed the work on Gasholders London, a development of 145 apartments within a triplet of listed gasholder guide frames.

King's Cross is the largest urban redevelopment scheme in Europe, with a rich industrial heritage integral to its renaissance. Among the most distinctive features to be retained was a triplet of Grade II-listed, cast iron, gasholder guide frames built in 1867. The triplet was abandoned as heavy industry moved to the outskirts of the city, and was dismantled in 2001 to allow for the Channel Tunnel Rail Link. The guide frames, including 123 columns, were painstakingly restored. Despite being over 150 years old, they were in remarkably good condition, largely preserved from decay by 32 layers of historic paint.

According to Founding Director Chris Wilkinson, the intention is to retain the presence of the structure but give it new meaning and use for the future. WilkinsonEyre's winning concept is housing the three residential buildings within the elegant frames. The design proposes three drums of accommodation at differing heights to suggest the movement of the original gasholders, which will have risen or

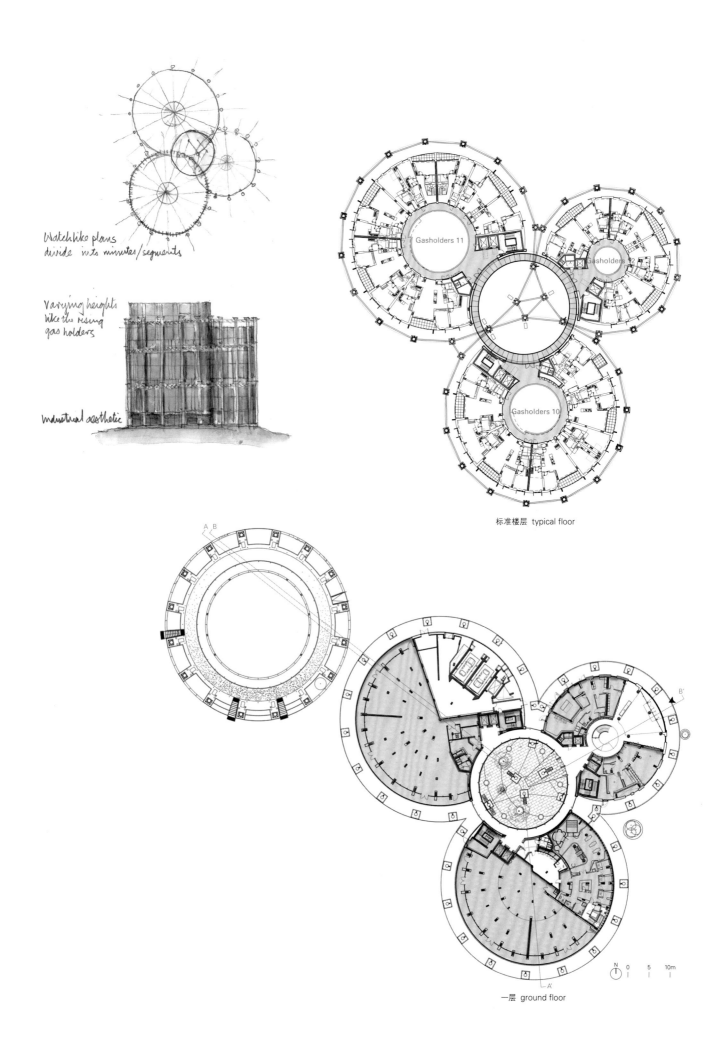

fallen depending on the pressure of the gas within. A fourth central virtual drum shape forms an open courtyard at the intersection of the cast iron structures. The design for Gasholders is developed to create a dynamic counterpoint between old and new. The heavy industrial aesthetic and raw physical materiality of the guide frames contrast with the lightness and intricacy of the interior spaces.

The scheme provides 145 apartments, a private gym and spa, a business lounge and an entertainment suite. Apartments are accessed through a central courtyard, each drum volume with its own atrium and core. These are linked by a series of circular walkways which surround the courtyard, where light is reflected in a central water feature. In another play of contrasts, the roofs are planted as gardens to bring nature to this re-inhabited urban landscape.

As Wilkinson said: "Working with circular geometry has resulted in really beautiful ideas. What began as a challenge, turned out to be a blessing." The circular nature of the buildings results in apartments that are laid out to take advantage of natural daylighting, with the living and bedrooms at the perimeter. This "pie" shaped configuration forms open-plan apartments with expansive views and a variety of orientations.

The cladding is composed of modular, vertical steel and glass panels, veiled by electronically-controlled shutters which provide occupants with shade and privacy. Designed for the busy urbanite, the apartments incorporate innovative technologies which will allow residents to adjust, control and environmentally fine-tune their living spaces.

This new vision for the gasholders delivers the functional requirements of modern living, whilst celebrating the character and form of the triplet guide frames. The project responds creatively to industrial heritage, meeting inherent constraints with imaginative design.

A-A' 剖面图 section A-A'

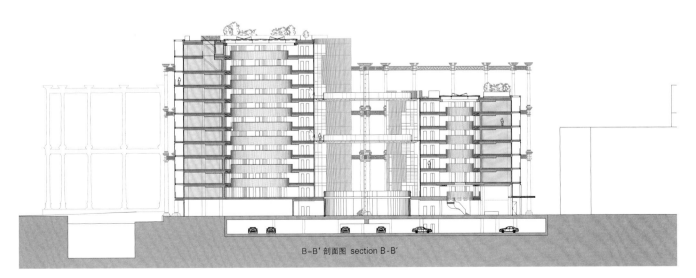

B-B' 剖面图 section B-B'

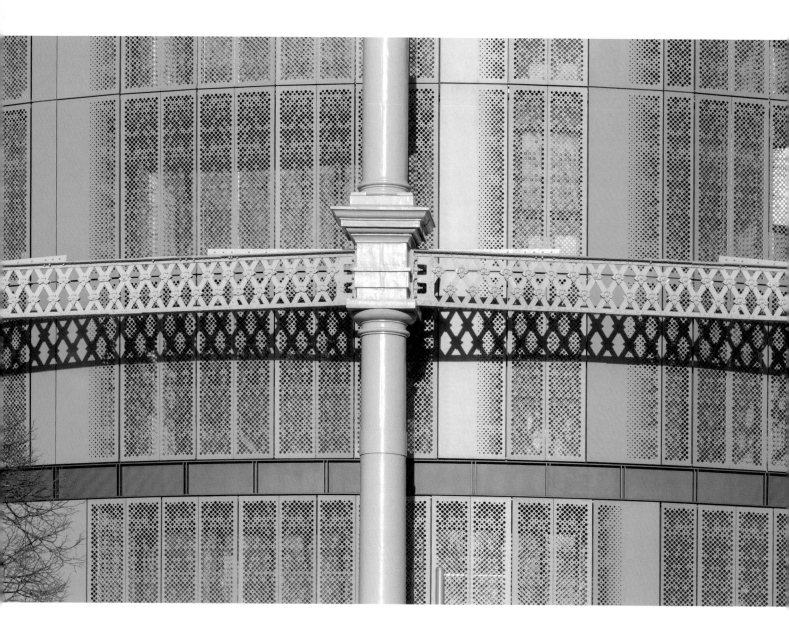

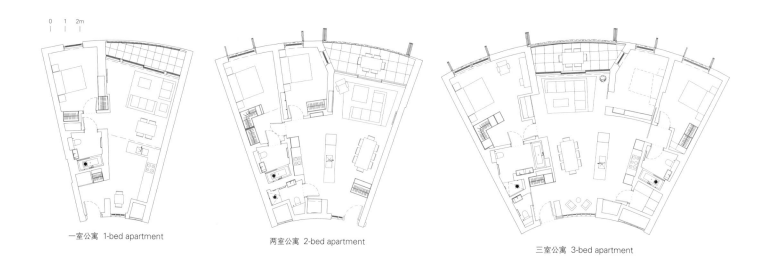

一室公寓 1-bed apartment 两室公寓 2-bed apartment 三室公寓 3-bed apartment

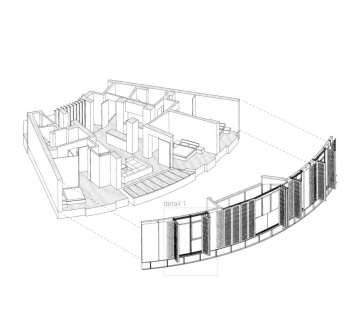

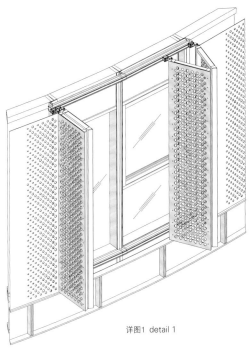

详图1 detail 1

项目名称：Gasholders London / 地点：London, UK / 事务所：WilkinsonEyre / 总承包商：Carillion / 住宅代理：Knight Frank / 项目管理、工料测量师：Gardiner & Theobald / 结构与立面工程师：Arup / 公寓室内建筑师：Jonathan Tuckey Design / 公共区域装修：No 12 Studio / 室外立面：Frener Reifer / 景观建筑师：Dan Pearson Studio / 规划顾问：NLP Nathan Lichfield & Partners / 照明顾问：Spiers & Major / 公共关系：Symposium International / 标志设计：Holmes Wood / 储气罐框架工程师：Craddy Pitchers Davidson / 消防、工程环境、可持续性、BREEAM、声学顾问、立面顾问、物流、运输、机电管道：Hoare Lea / CDM协调：David M Eagle 规格制定：Aecom / 客户：King's Cross Central Limited Partnership / 用途：residential / 总楼面面积：23,950m² / 单元数量：17 x Studio, 34 x 1 Bed, 43 x 2 Bed, 4 x 2 Bed Duplex, 34 x 3 Bed, 4 x 3 Bed Duplex, 9 x Penthouses, private gym and spa, business lounge, entertainment suite and screening room, private bar and dining room / 结构：reinforced concrete frame throughout, original gasholder frames made of cast iron / 外围护结构：perforated metal rain screen on a unitised glazing system to exterior with sliding folding shutter system throughout, ultra-high strength precast concrete cladding to the courtyard / 阳台：steel bolt on balconies behind sliding folding shutters / 设计时间：2002—2013 / 施工开始时间：2015.2 / 竣工时间：2018.1 / 摄影师：©Peter Landers (courtesy of the architect) - p.197, p.200, p.201, p.202^{lower}, p.204, p.205; ©John Sturrock (courtesy of the architect) - p.202^{upper}, p.203^{lower}; courtesy of the architect - p.194~195

Novetredici 住宅综合楼
Novetredici Residential Complex
Cino Zucchi Architetti

一层 ground floor

坐落于米兰via De Cristoforis大街上的Novetredici住宅综合楼是由Cino Zucchi Architetti建筑事务所设计的，它的建成标志着via Viganò大街、Rosales大街和de Cristoforis大街以及Fratelli Castiglioni大街所围成的街区的完整，Fratelli Castiglioni大街的旁边矗立着巴勒莫新门开发项目的新Unicredit摩天大楼综合建筑。

Novetredici大楼的北侧面对的是新的城市改造工程，南侧面对的则是原有的城市结构，主要是由周边的建筑组成的。它们是"过渡"性的元素，如via Viganò大街上的原有住宅，这些住宅都有面向街道开放的庭院，还有第二次世界大战后的一些建筑，其布局更为自由，拥有更高的高度。

Novetredici大楼与CZA附近的综合设施La Corte Verde一起，修改了该地区的总体规划，创造了一种居住模式，该模式强化了原有的城市结构，同时也创造出新的环境质量。这两个建筑群不需重新创建周边建筑，自身的布局就具有强烈的城市特征：新的建筑三面与原有道路对齐，而西面则保持开放状态，使得绿色的中心地带和via Rosales大街能够进行视觉上的交流。由总体规划（一个高高的、线性的建筑体，可以在花园投射出长长的阴影，这样就在北面创造出一块永远没有阳光的区域）确定的建筑外围护结构被分为两部分，一部分是较低的西部，另一部分是较高的东部，两部分在底部通过一个共同的入口大厅连接。两部分采用集中布局的方式，设计既高效又紧凑，注意到了原有城市边缘与太阳运行规律的关系，并在实心墙和大型阶梯之间产生了一种柔和的递进的感觉，它们又构成了居住环境的遮挡。

西侧的阶梯式建筑与科索科莫的历史建筑相呼应。在东面，它增加了via Viganò大街的宽度，因而可以更深入地感受历史名城和巴勒莫新门开发项目。

公共中庭的破碎式几何结构，以立于细长圆柱上的薄石板和玻璃营造的周边为标志，创造了一个城市化的门厅，同时拓宽了原先狭窄的via De Cristoforis大街，在街道和后面的私人花园之间创造了透明感，同时又与La Corte Verde产生了模糊的联系感。

通往私人地下停车场的通道位于建筑体量之内；两个主体的底部以大块做旧的白色石头为标志，与原有街道在视线水平上产生了形式上的延续性。在这一点上，玻璃护墙强有力的水平线平衡了金属元素和窗户的垂直性。各种颜色和表面纹理在非常单一的设计中产生了连续变化的感觉。

两座建筑物的单坡屋顶向中央花园汇聚。它们倾斜的外观强化了住宅建筑群的城市特征，能够将完善的建筑主旨与那些在公共空间漫步的人的多元化的感知状态结合起来。尽管Novetredici住宅综合楼以不同的材料和建筑解决方案为特点，但它与La Corte Verde形成了强烈的呼应和一系列形式上的共鸣，为"内城生活"的主题赋予了新的尊严，并在摈弃参考和怀旧的基础上，重新诠释了第二次世界大战后在米兰最好的住宅建筑中所体现出来的"现代传统"。

Residential Complex Novetredici in via De Cristoforis, Milan, designed by via Cino Zucchi Architetti, represents the natural completion of the block bordered by via Viganò, Rosales, de Cristoforis and Fratelli Castiglioni streets; the latter flanks the new Unicredit skyscraper complex in the Porta Nuova development.

The Novetredici block hinges the new urban transformation on its north side and the existing urban fabric on the south, which is formed by perimeter buildings, "transitional" elements – such as the existing houses in via Viganò, with open courtyards onto the street – and by post-WWII architecture with its freer disposition and greater height.

Along with CZA's nearby complex, La Corte Verde, Novetredici revises the masterplan guidelines for the area in favor of a settlement model which reinforces the existing urban structure whilst also generating a new environmental quality. The arrangement of the two complexes is marked by a strong urban character without necessarily recreating a perimeter block: the new volumes align on three sides with existing roads, whereas the west side is left open, allowing the green heart and via Rosales to communicate visually. The building envelope defined by the masterplan – a tall,

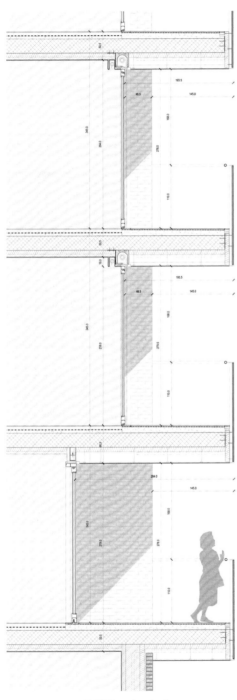

剖面详图 section detail

立面详图 elevation detail

项目名称：Novetredici Residential Complex / 地点：Milano, Italy / 事务所：Cino Zucchi Architetti / 项目主管：Cino Zucchi, Andrea Viganò, Barbara Soro, Valentina Zanoni
合作者：con Andrea Balestreri, Anna Braghini, Alberto Brezigia, Giulia Buzzoni, Omar de Ciuceis, Chiara Toscani / 结构顾问：Progetti e Strutture – F. Scarantino, G.M. Iselle
施工顾问：Ai Studio - Gian Paolo Bottan, Enrico Fabris, Giorgio Macri / 执行公司：Italiana Costruzioni / 用地面积：2,199m² / 总楼面面积：6,690m² / 建筑造价：4,500,000 €
竣工时间：2011 / 施工时间：2014—2016 / 摄影师：©Filippo Poli (courtesy of the architect)

linear, body that would have cast its long shadow over the garden and created a north side perpetually without sun – is instead divided into two, a lower western part and a higher eastern one, joined at the base by a common entrance hall. Two volumes combine the efficiency and compactness of a centralized plan, paying attention to the relationship between the existing urban edges and the course of the sun, and producing a soft progression between solid walls and large terraces that together constitute an inhabited screen.

The stepped profile on the west side is prescribed by proximity to historical buildings in Corso Como. To the east it enhances the width of via Viganò, with long views to the historic city and the new Porta Nuova development.

The broken geometry of the common atrium, marked by a thin slab resting on slender cylindrical pillars and a glazed perimeter, creates an urban foyer which expands the narrow via De Cristoforis, creating transparency between the street and the rear private gardens which blur with La Corte Verde. Access to a private underground car park is located within the building volume; the base of both bodies is marked by large slabs of rusticated white stone, creating a formal continuity at eye level along the existing street. Over this, the strong horizontal lines of glass parapets balance the verticality of metallic elements and of the windows. A variety of colors and surface textures generate continuous variation within a strongly unified design.

The single-pitched roofs of both buildings converge towards the central garden. Their slanted profiles reinforce the urban character of a housing complex capable of combining a sound architectural statement with the plurality of perceptual states for those who stroll through public spaces. Although marked by different materials and architectural solutions, the Novetredici complex creates a strong dialogue with La Corte Verde, creating together a series of formal resonances which give new dignity to the theme of "inner city living", and reinterpret the "modern tradition" of the best post-WWII residential architecture in Milan without direct reference or nostalgia.

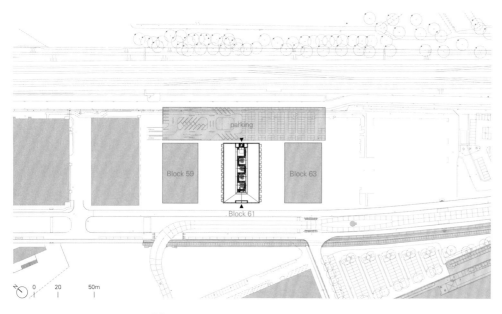

Loftwonen 61 号楼
Loftwonen Block 61
Architecten|en|en

从几年前开始，位于荷兰工业城市埃因霍温的飞利浦电子公司的前工业园Strijp-S就在进行一次大规模的重新开发。Strijp-S的城市设计强调了飞利浦传统建筑特征和定制新建筑的结合。团队的驱动力是创造力，创造性的职业和功能要与城市生活环境相结合。

Loftwonen是这种环境下城市居住的一种特殊形式。Loftwonen的理念基于高举架的房间和灵活的楼层布局。61号楼就是以类似方式建造的三座住宅楼的城市组合体的其中之一。这三座住宅楼位于铁路沿线，是著名活动区Klokgebouw的延伸，但通过一个与铁路路基平行的长方形停车场与铁路分隔开。

新住宅楼无缝地融入了以大胆的工业建筑为特色的街区。在大量重复的模式中使用的红砖立面和预制的彩色混凝土创造了一个可以匹配其周围环境的坚固的建筑。

这座七层公寓楼由环绕着一个大型共用庭院的两翼组成。在这一区域，人们可以找到通向住宅单元的入口和一部所谓的"瀑布式楼梯"。与坚固的外观相反，庭院的设计采用了特殊的小规模元素。这个院子里的一切都是天蓝的色，在充满活力的Strijp-S区域创造了一片宁静的绿洲。在建筑的基座结构中，入口附近设有许多商业空间。

房间配备了卫生设施，也被称为"小房间"，包括厨房、卫生间和浴室，这样可以根据需要灵活地安排其余空间，加上公寓不寻常的3.20m的举架高度，增加了每个住宅单元的空间自由度。

Since a couple of years ago, Strijp-S, the former industrial park that belonged to an electronics company Philips, in the Dutch industrial city of Eindhoven, has been undergoing a major redevelopment. The urban design of Strijp-S emphasizes a combination of the Philips' heritage and customized new buildings. The driving force of the ensemble was creativity; creative professions and functions mixed with an urban living environment.

Loftwonen is a special form of urban residence in this setting. The idea of Loftwonen is based on high-ceiling rooms and flexible floor plans. Block 61 is a part of an urban composition of the three housing blocks that are constructed in a similar manner. The housing blocks are located along the railway as an extension of the well-known event area, Klokgebouw, but are separated from the railway by an oblong parking garage, which lies parallel to the railway embankment.

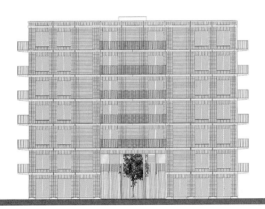

西南立面 south-west elevation

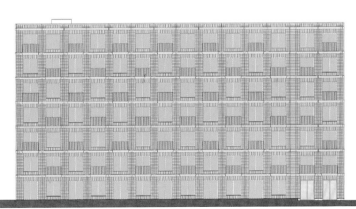

东南立面 south-east elevation

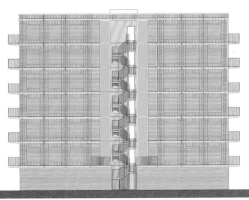

东北立面 north-east elevation

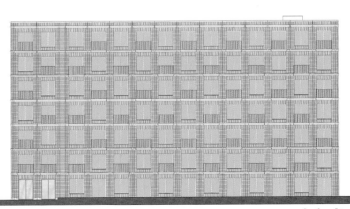

西北立面 north-west elevation

The architecture of the new housing blocks seamlessly merges into the neighborhood characterized by bold industrial buildings. Facades constructed from red brickwork and prefabricated colored concrete in massive repetitive patterns create a robust architecture that matches its surroundings.

The seven-layer apartment complex consists of two wings that surround a large shared courtyard. In this area, one can find the galleries with the entrances to the housing units and a so-called "waterfall stairway". Contrary to its robust exterior appearance, the courtyard has been designed with small-scale elements in a special manner. Everything in this courtyard is colored in sky blue, creating a peaceful oasis within the dynamic Strijp-S area. In the plinth of the building, a number of commercial space can be found next to the entrance.

The rooms have been equipped with sanitation blocks, also known as "cubbies". A "cubby" includes a kitchen, a toilet, and a bathroom. This allows the rest of the space to be arranged flexibly as required. Together with the apartments' unusual floor heights of 3.20m, it adds to spatial freedom inside each housing unit.

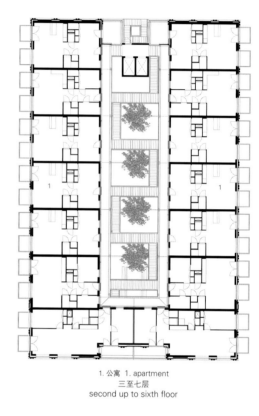

1. 公寓　1. apartment
三至七层
second up to sixth floor

1. 大门　2. 庭院　3. 公寓　4. 商业空间
1. gate　2. courtyard　3. apartment　4. commercial space
一层　ground floor

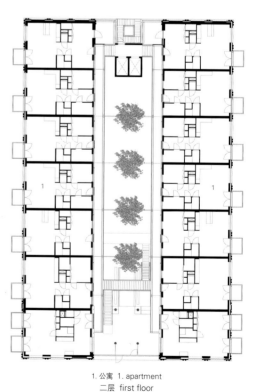

1. 公寓　1. apartment
二层　first floor

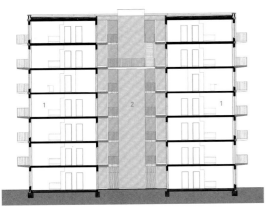

1. 公寓 2. 庭院　1. apartment 2. courtyard
A-A' 剖面图　section A-A'

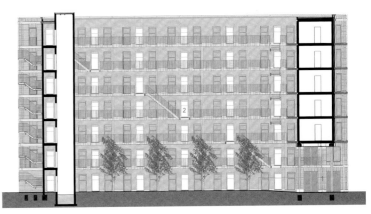

1. 大门 2. 庭院　1. gate 2. courtyard
B-B' 剖面图　section B-B'

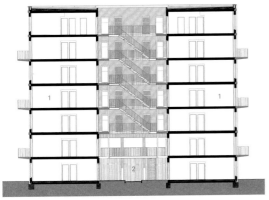

1. 公寓 2. 庭院　1. apartment 2. courtyard
C-C' 剖面图　section C-C'

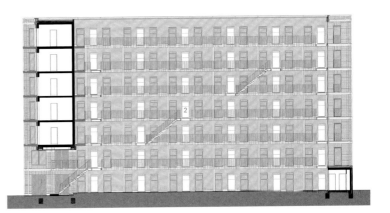

1. 大门 2. 庭院　1. gate 2. courtyard
D-D' 剖面图　section D-D'

项目名称：Loftwonen Block 61 / 地点：Eindhoven, The Netherlands / 事务所：Architecten|en|en / 项目团队：Arie van Rangelrooij, Martijn Wilms, Pim van der Linden, Philemon Henselmans / 承包商：Stam + De Koning / 建议：Peutz / 施工：Aveco De Bondt / 设备：Homij / 客户：SDK Vastgoed
用途：96 apartments (rental), commercial space / 用地面积：2,550m² / 建筑面积：1,055m² / 总楼面面积：7,330m² / 设计时间：2013 / 竣工时间：2017
摄影师：©BASE Photography (courtesy of the architect)

P206　Cino Zucchi Architetti
As one of the leading Italian studios in contemporary architecture, is constantly searching new spatial solutions in the rapidly changing context of urban landscapes. Born in Milano in 1955, Cino Zucchi[p.229, left-bottom] earned a Bachelor of Science in Art and Design at the M.I.T. in 1978 and a Laurea in Architettura at the Politecnico di Milano in 1979, where is currently Chair Professor of Architectural and Urban Design. Has been the curator of the Italian Pavilion at the 2014 Venice Biennale of Architecture. Has been the president of the Jury of the Mies van der Rohe Award 2015.

P32　CODE: Architecture AS
Was established in 1997 in Oslo, Norway. Currently has 12 employees and five partners including Bjarne Ringstad[p.229, right-bottom], Ole Einejord, Gaute Stensrud, Julian Lynghjem, and Henning Kaland. Has won a number of national and international architecture competitions. Its portfolio includes residences, commercial buildings, schools, hotels, landscape and a number of urban development projects. Many projects are carried out with high ambitions regarding sustainability and energy efficient solutions.

P70　Haworth Tompkins
Is a British architectural studio based in London, voted Building Design Architect of the Year and RIBA London Architect of the Year in 2014. Founded in 1991 by Graham Haworth (1960) and Steve Tompkins (1959), the studio has an international reputation for theater design, the Liverpool Everyman Theater winning the 2014 RIBA Stirling Prize for the best building of the year by a UK architect. Was part of the Gold Award UK winning team at the Prague Quadrennial and was chosen to exhibit theater work at the 2012 Venice Biennale.

©Rob Greig

P194　WilkinsonEyre Architects
Is one of the UK's leading architectural practices, based in London since 1999. Was twice awarded both the prestigious RIBA Stirling Prize and the RIBA Lubetkin Prize. Chris Wilkinson[picture-above] founded Chris Wilkinson Architects in 1983 and formed a partnership with Jim Eyre in 1987. His contribution has been recognised by the award of an OBE in the Millennium Honours List, election to the Royal Academy of Arts in 2006, and an Honorary Fellowship of the American Institute of Architecture in 2007. He combines a lifelong interest in art with a fascination for science, technological innovation and a sense of history, producing a fresh new approach to architecture.

©Brigitte Lacombe

P46 Atelier Peter Zumthor

Peter Zumthor was born in Basel, Switzerland in 1943. After studying at the Kunstgewerbeschule(School for Arts and Crafts) in Basel, he studied Industrial Design and Architecture as an exchange student at Pratt Institute in New York. Founded his own firm in 1978 in Haldenstein, Switzerland. Has taught at the Accademia di Architettura di Mendrisio from 1996 to 2008. Also has been a visiting professor at SCI-ARC, the Technical University of Munich, Tulane University, and the Harvard Graduate School of Design. Was elected to the Academy of Arts, Berlin in 1994. Was made an honorary member of the Bund Deutscher Architekten (BDA) in 1996. Received the Carlsberg Architectural Prize (1998), Mies van der Rohe Award (1999), Praemium Imperiale (2008), Pritzker Architecture Prize (2009), RIBA Royal Gold Medal (2013).

P04 Silvio Carta

Dr. Silvio Carta is an ARB RIBA architect and head of Design and Visual Arts, and chair of the Design Research Group at the University of Hertfordshire. His studies have focused on digital design, digital manufacturing, design informatics, data visualisation and computational optimisation of the design process. Silvio is an editor of A_MPS Architecture Media Politics and Society (UCL Press), and the curator of the international lecture series AUDITORIUM 2015-16: The Architecture of Information, Data, People and Public Space (Leuven, Belgium). Is head of the editorial board of C3 magazine, Korea.

P158 Heidi Saarinen

Is a London based designer, lecturer at Coventry University and also an artist with current research focused on space and place. Is interested in the peripheral space, in-between and the interaction and collision between architecture, spaces, city, performance and the body. Is currently working on a series of interdisciplinary projects linking architecture, heritage, film and choreography in the urban environment. Is part of several community and c reative groups in London and the UK where she engages in events and projects highlighting awareness of community and architectural conservation in the built environment.

P178 BIG

Founded in 2005 by Bjarke Ingels, BIG is a Copenhagen, New York and London based group of architects, designers, urbanists, landscape professionals, interior and product designers, researchers, and inventors. Currently involves in a large number of projects throughout Europe, North America, Asia, and the Middle East. Believes that in order to deal with today's challenges, architecture can profitably move into a field that has been largely unexplored. A pragmatic utopian architecture that steers clear of the petrifying pragmatism of boring boxes and the naïve utopian ideas of digital formalism. Like a form of programmatic alchemist, it creates architecture by mixing conventional ingredients such as living, leisure, working, parking, and shopping. By hitting the fertile overlap between pragmatic and utopia, once again finds the freedom to change the surface of our planet, to better fit contemporary life forms.

P24 Jarmund / Vigsnæs AS Arkitekter MNAL

Was established in 1996 in Oslo, Norway by Einar Jarmund[p.231, upper picture, left] and Håkon Vigsnæs[right]. Alessandra Kosberg[center] has joined as a partner in 2004. Einar and Håkon were born in Oslo, 1962 and graduated from the Oslo School of Architecture. Alessandra was born in 1967, graduated from the Oslo School of Architecture in 1995 and started working with JVA in 1997. After studying at the AA School in London, Håkon worked with Sverre Fehn. Einar taught and worked in Seattle after getting a Master's degree from University of Washington in Seattle. Both were visiting Professors at Washington University, University of Arizona and Rhode Island School of Design. Håkon has been teaching at their alma mater since 2013.

P24

P214 Architecten|en|en

Senior Architect, Arie van Rangelrooy[second-left] graduated in 1904 from the Eindhoven University of Technology (TU/e), Department of Architectural History & Theory. Joined forces in 1987 with Hans Thomassen, where he became co-owner in 1991. Robbert Urlings[right] is office manager and finance specialist. Graduated in 1993 from the Eindhoven University of Technology, Department of Architectural and Urban Design. Pim van der Linden[fourth-left] is chef de bureau and IT specialist. Graduated in 1998 at the Eindhoven University of Technology, Department of Architectural Design. Is responsible for the internal coordination and the technical quality of the drawing work. Robbert and Pim have been working for Architecten|en|en since 1995. Senior Architects, Frans Benjamins[left] & Joost Verbeek[third-left] studied building technology at Hogeschool 's-Hertogenbosch and graduated with honors as Master of Architecture from the Tilburg Academy for Architecture & Urbanism.

P12 Ghilardi+Hellsten Arkitekter AS

Is an Oslo based practice founded in 2005 by Franco Ghilardi and Ellen Hellsten. The office is engaged in architecture, landscape and urbanism as a common discipline. Most of its commissions have been acquired through open or restricted competitions. Its works are renowned for operating on a non-dogmatic and alternative ground along the traditional Scandinavian design. Provides architectural and planning services on all levels, from concept design to design development, detailing, construction supervision and approval process management. The office is a member of the National Association of Norwegian Architects (NAL), The Association of Consulting Architects in Norway and the Norwegian Green Building Council. Eldhusøya tourist route project has been nominated for the German Design Award 2017 and shortlisted for the 2017 Mies van der Rohe Award.

©BASE Photography

©Marcus Hawk

P118 Heatherwick Studio
Founded by Thomas Heatherwick[picture-above] in 1994, is a team of 250 problem solvers dedicated to making the physical world around us better for everyone. Based out of its combined workshop and design studio in Central London, it creates buildings, spaces, master-plans, objects and infrastructure. Thomas Heatherwick is a British designer whose prolific and varied work over two decades is characterised by its ingenuity, inventiveness and originality. The studio recently completed Coal Drops Yard. Lisa Finlay, since joining Heatherwick Studio in 2011 as Group Leader, worked closely with the developer, heritage bodies and stakeholders to introduce a curving roof to stitch the historic coal drops buildings together – creating a third storey of retail and unifying heart to the site. Tamsin Green joined the studio in 2012 and took on the role of Project Leader on Coal Drops Yard.

P06 Per Ritzler
Is senior adviser at the Norwegian Scenic Routes in Norway, with responsibility for national and international promotion of the Scenic Routes. Is a former journalist and correspondent for the Swedish national broadcasting corporation, and has also previously run his own consulting company dealing with communication and journalism.

P62 Isabel Potworowski
Graduated from TU Delft with a Master in Architecture, and currently works for Barcode Architects in Rotterdam. During the graduate studies, Potworowski was a member of the editorial committee and wrote several articles for the independent student journal *Pantheon*. Originally from Canada, she completed her Bachelor in Architecture at McGill University in Montreal, where she was awarded the Louis Robertson book prize for the highest grade in architectural history. Has also studied one semester at the Politecnico di Milano. Has worked at ONPA Architects and Manasc Isaac Architects, both in Edmonton, Canada.

P100 van Dongen-Koschuch
Was co-founded in 2012 by Frits van Dongen[right] and Patrick Koschuch[left], members of the Royal Institute of Dutch Architects. Frits van Dongen graduated from the TU Delft in 1980. Established Van Dongen Architekten in Delft in 1985 and co-founded 'de Architekten Cie.' in 1988, in which he has been a partner till 2012. Received BNA Kubus Award in 2006. Was Chief Government Architect of the Netherlands from 2011 to 2015. Is a member of the Dutch Professional Organisation of Urban Designers and Planners. Patrick Koschuch, born in Vienna, Austria, studied Economical Geography at the Maximilian University and graduated from the TU Delft in 1998. Worked at de Architekten Cie. as an Associate Architect till 2012 and ran his own studio Patrick Koschuch Architecten from 2006 until 2012.

P144 NIO Architecten

Dutch architect and writer, Maurice Nio (1959)[p.233, upper] graduated in 1988 from the Faculty of Architecture of TU Delft, his thesis project being a villa for Michael Jackson: the most peculiar thesis project that year. This project has been of vital importance to his hybrid approach. Founded the unconventional design agency, NOX in the early '90s with the architect Lars Spuybroek. Founded the Rotterdam-based Design Studio NIO Architecten together with Joan Almekinders in 2000.

P84 KAAN Architecten

Was founded at Rotterdam by Kees Kaan, Vincent Panhuysen and Dikkie Scipio in 2014. Has been operating São Paulo branch since 2015, led by Renata Gilio.
Kees Kaan was the founding partner of Claus en Kaan Architecten. Graduated at the Faculty of Architecture of TU Delft in 1987. Has been professor of Architectural Design at TU Delft since 2006. Vincent Panhuysen joined Claus en Kaan Architecten in 1997 and became partner in 2002. Studied at the Utrecht School of the Arts(HKU) and the Rotterdam Academy of Architecture. Dikkie Scipio joined Claus en Kaan Architecten in 1995 and became partner in 2002. Studied at the Royal Academy of Art, The Hague and the Rotterdam Academy of Architecture.

P166 Reinier de Graaf / OMA

Was born in 1964 in Schiedam, the Netherlands and graduated from Stedelijk Gymnasium Schiedam in 1982. Holds an architecture diploma from TU Delft and a master's degree in architecture from the Berlage Institute. Joined OMA in 1996, he is a partner of Office for Metropolitan Architecture (OMA), where he leads projects in Europe, Russia and the Middle East. In 2002, he co-founded AMO, the think tank of OMA. His recent built work includes the Timmerhuis, a mixed-use project in Rotterdam; fashion brand G-Star Raw's corporate and design headquarters in Amsterdam; and De Rotterdam, currently the largest building in the Netherlands.

P134 COBE

Was founded in Copenhagen, Denmark by Danish architect, Dan Stubbergaard[p.233, lower picture] in 2005. Employs around 100 dedicated architects, constructing architects, landscape architects and urban planners. Has been particularly successful within the area of urban planning with a number of award-winning designs of buildings and urban spaces. Received the prestigious Golden Lion Award at the Venice Biennale in 2006, Nykredit Architecture Prize in 2012, Dreyer's Foundation's Honorary Award in 2015 and the Eckersberg Medal in 2016.

© 2020 大连理工大学出版社

版权所有·侵权必究

图书在版编目(CIP)数据

新旧熔融：汉英对照 / (英) 赫斯维克建筑事务所等编；曹麟，邵磊译. — 大连 : 大连理工大学出版社，2020.5
(建筑立场系列丛书)
ISBN 978-7-5685-2522-0

Ⅰ. ①新… Ⅱ. ①赫… ②曹… ③邵… Ⅲ. ①旧建筑物—旧房改造—建筑设计—汉、英 Ⅳ. ①TU746.3

中国版本图书馆CIP数据核字(2020)第063886号

出版发行：大连理工大学出版社
　　　　　（地址：大连市软件园路80号　邮编：116023）
印　　刷：上海锦良印刷厂有限公司
幅面尺寸：225mm×300mm
印　　张：14.25
出版时间：2020年5月第1版
印刷时间：2020年5月第1次印刷
出 版 人：金英伟
统　　筹：房　磊
责任编辑：杨　丹
封面设计：王志峰
责任校对：张昕焱
书　　号：978-7-5685-2522-0
定　　价：298.00元

发　行：0411-84708842
传　真：0411-84701466
E-mail：12282980@qq.com
URL：http://dutp.dlut.edu.cn

本书如有印装质量问题，请与我社发行部联系更换。

墙体设计
ISBN: 978-7-5611-6353-5
定价: 150.00元

新公共空间与私人住宅
ISBN: 978-7-5611-6354-2
定价: 150.00元

住宅设计
ISBN: 978-7-5611-6352-8
定价: 150.00元

文化与公共建筑
ISBN: 978-7-5611-6746-5
定价: 160.00元

城市扩建的四种手法
ISBN: 978-7-5611-6776-2
定价: 180.00元

复杂性与装饰风格的回归
ISBN: 978-7-5611-6828-8
定价: 180.00元

内在丰富性建筑
ISBN: 978-7-5611-7444-9
定价: 228.00元

建筑谱系传承
ISBN: 978-7-5611-7461-6
定价: 228.00元

伴绿而生的建筑
ISBN: 978-7-5611-7548-4
定价: 228.00元

微工作·微空间
ISBN: 978-7-5611-8255-0
定价: 228.00元

居住的流变
ISBN: 978-7-5611-8328-1
定价: 228.00元

本土现代化
ISBN: 978-7-5611-8380-9
定价: 228.00元

都市与社区
ISBN: 978-7-5611-9365-5
定价: 228.00元

木建筑再生
ISBN: 978-7-5611-9366-2
定价: 228.00元

休闲小筑
ISBN: 978-7-5611-9452-2
定价: 228.00元

景观与建筑
ISBN: 978-7-5611-9884-1
定价: 228.00元

地域文脉与大学建筑
ISBN: 978-7-5611-9885-8
定价: 228.00元

办公室景观
ISBN: 978-7-5685-0134-7
定价: 228.00元